ASTRONOMY AND MATHEMATICS IN ANCIENT INDIA
ASTRONOMIE ET MATHÉMATIQUES DE L'INDE ANCIENNE

LETTRES ORIENTALES ET CLASSIQUES

Collection dirigée par Sylvie Vanséveren (Université Libre de Bruxelles)

La collection *Lettres Orientales et Classiques* est dédiée aux langues, aux textes et aux civilisations qui composent le monde méditerranéen antique aussi bien que les civilisations de langue indo-européenne. Elle a pour objectif de publier des travaux et des ouvrages de référence consacrés aux langues, à la culture, la mythologie, la religion dans une optique scientifique qui allie comparaison et reconstruction, diachronie et histoire, recherche et pédagogie.

LETTRES ORIENTALES ET CLASSIQUES 17

ASTRONOMY AND MATHEMATICS IN ANCIENT INDIA

ASTRONOMIE ET MATHÉMATIQUES DE L'INDE ANCIENNE

Actes de la journée d'études organisée le 24 avril 2009 à l'Université Libre de Bruxelles

édités par
J.M. DELIRE

PEETERS
LEUVEN – PARIS – WALPOLE, MA
2012

Illustration de la couverture: L'astrolabe pour Dōmodara, détail,
photographie de Marie Levaux.

ISBN 978-90-429-2614-1 (Peeters Leuven)
D/2012/0602/74

Table des matières

K. Ramasubramanian lors du colloque

Préface

Si l'on compare l'évolution du nombre de cours (en formation initiale ou continue), de mémoires ou de thèses consacrés à l'histoire des sciences en Belgique aux mêmes données pour les pays voisins, particulièrement la France, on ne peut que constater un retard de cette discipline dans notre pays malgré l'intérêt pédagogique, de plus en plus proclamé, qu'elle présente pour l'enseignement des sciences. Par exemple, à l'ULB – que nous remercions ici pour avoir hébergé cette journée d'étude à titre gracieux – ont été mis sous le boisseau ces dernières années pas moins de trois cours consacrés à l'histoire des sciences et des techniques. Malgré cela, la discipline continue à susciter des vocations et des activités, que leurs initiateurs réalisent tant bien que mal dans cette ambiance générale peu encourageante. On peut citer, toujours à l'ULB, le Centre Altaïr d'histoire des sciences et des techniques, une association reconnue par le Conseil de la Recherche qui organise depuis plus de 20 ans, et sans aucun subside, des cycles de conférences et des journées d'études dont le succès atteste de l'intérêt du public pour ces sujets.

Ceci dit, même dans les pays proches qui développent ou maintiennent une tradition forte, l'étude de l'histoire des sciences « non occidentales » reste le parent pauvre. L'accent, mis depuis quelques décennies sur la construction de l'Europe y est peut-être pour quelque chose. Pourtant, que seraient l'astronomie et les mathématiques aujourd'hui sans l'apport des savants arabo-musulmans, eux-mêmes grands amateurs des textes indiens et grecs – sans parler des chinois – qu'ils ont traduits et commentés avant de les transmettre à l'Europe ? Le plus grand ouvrage grec d'astronomie, l'*Almageste*, est connu sous le nom que lui donnaient les savants arabes. Les algorithmes, si utiles en informatique, doivent leur appellation au premier mathématicien connu d'expression arabe, Al-Khwarīzmī, dont le nom atteste qu'il était originaire d'une région proche de l'Inde, où il a puisé son savoir. Les Grecs anciens eux-mêmes, que l'on veut à toute force placer à l'origine de la science « occidentale », reconnaissent largement leur dette à l'Orient.

La connaissance que nous avons aujourd'hui des sciences indiennes et arabo-musulmanes est parcellaire, tributaire du nombre de manuscrits recensés et édités. Pour l'Inde, moins de 10% des ouvrages scientifiques sanskrits ont été édités, si l'on en croit K. V. Sarma (*« Sanskrit and Science : A New Area of Study », Papers of the 12th World Sanskrit Conference*, vol.7, Delhi, 2009, p.10), qui a consacré sa vie entière (1919-2005) à l'édition de manuscrits astronomiques indiens. Le peu que nous savons montre déjà que les savants indiens ont contribué de manière non négligeable à la connaissance scientifique mondiale. Dans le domaine de l'astronomie et des mathématiques, sujet de cette journée d'étude, ils ont commencé très tôt, puisque le rituel védique (premier millénaire av. J.C.) nécessitait déjà des connaissances élémentaires en géométrie, arithmétique et astronomie. Dès le début de l'ère chrétienne, alors qu'était mis au point le système de numération décimal (que nous devons à l'Inde, faut-il le rappeler), apparaissaient les siddhāntas, traités d'astronomie mathématique dont certains utilisaient déjà la notion de sinus (une autre innovation que nous devons à l'Inde). Puis vint Āryabhaṭa, le premier astronome-mathématicien original connu de nom, et cette originalité ne se démentit pas pendant les douze siècles suivants, jusqu'à ce que les Anglais imposent dans leurs collèges anglo-indiens l'enseignement de la science européenne. Les différents articles de ce volume éclairent certains aspects de cette contribution indienne originale.

R. Mercier (*The Reality of Indian Astronomy*) reprend les travaux de Billard sur l'origine des paramètres utilisés par Āryabhaṭa (Ve siècle) pour calculer les positions moyennes des planètes à partir du nombre de jours (ahargaṇa) écoulés depuis le début de son canon, fixé au 18 février -3101. Il montre que, contrairement à ce qu'affirmait David Pingree, les astronomes indiens ont noté des observations pour améliorer la précision des paramètres qu'ils utilisaient dans leurs canons (siddhāntas), même si – R. Mercier ne le nie pas – certains des ces paramètres sont d'origine grecque. Le rattachement des canons aux ères définies par la tradition ancienne de l'Inde (*Purāṇas*), en particulier le début du canon d'Āryabhaṭa identifié avec le début du Kaliyuga, ne doit pas être interprété comme signifiant que cette tradition serait à l'origine des siddhāntas. En particulier, il ne faut pas voir, comme le font David Pingree et sa disciple Kim Plofker, le Pitāmahāsiddhānta du *Viṣṇudharmottara Purāṇa* comme un réel siddhānta, puisque, étant incomplet, il ne permet aucun calcul.

K. Ramasubramanian (*Indian Planetary Models : Āryabhaṭa to Nīlakaṇṭha*) rappelle de manière très détaillée la procédure suivie par les astronomes indiens, depuis Āryabhaṭa, pour calculer la position vraie en longitude des planètes à partir de leurs positions moyennes, via les corrections maṇḍa et śīghra. Il montre ensuite, par des extraits de ses différents ouvrages, comment Nīlakaṇṭha, après une longue lignée d'astronomes kéralais, parvint, à l'aide de son interprétation géométrique de cette procédure, à expliquer le traitement différent appliqué aux planètes inférieures (Mercure, Vénus) et aux planètes extérieures. K. Ramasubramanian insiste beaucoup sur l'intérêt, pour ces considérations globales, du petit ouvrage tardif de Nīlakaṇṭha, le *Grahasphuṭānayane vikṣepavāsanā.* Enfin, il montre que Nīlakaṇṭha a été le premier astronome à rendre compte de la latitude (vikṣepa) des planètes d'une manière plus satisfaisante que Ptolémée et Copernic.

P. -S. Filliozat (*Mathématiques et scolastique dans l'Inde médiévale, l'exemple du Haricarita de Parameśvara Bhaṭṭa*) présente le Haricarita (XIVe siècle) comme un ouvrage poétique et un acte de dévotion viṣṇuite, puisqu'il résume un chapitre du *Bhāgavata Purāṇa* consacré à l'enfance de Kṛṣṇa. Mais le Haricarita est aussi un ouvrage d'astronomie qui s'impose pour début de ses 248 strophes les nombres exprimant les 248 longitudes de la lune durant 9 mois anomalistiques, exprimées dans une numération littérale (kaṭapayādi). Ces mots-nombres, les vākyas, furent les premières informations sur l'astronomie indienne recueillies par l'astronome Guillaume Le Gentil, venu à Pondichéry (1761-9) observer les passages de Vénus devant le soleil. Après avoir expliqué le fonctionnement astronomique de ces nombres et donné des arguments pour l'attribution du *Haricarita* à Parameśvara Bhaṭṭa (fin XIIIe, début XIVe siècles), P. -S. Filliozat montre quelques exemples d'insertion du vākya dans une strophe.

S. R. Sarma (*Yantrarāja for Dāmodara, The Earliest Extant Sanskrit Astrolabe*) est un grand connaisseur des astrolabes (yantrarāja) sanskrits, qu'il a répertoriés et dont il prépare un catalogue. Il présente ici le plus ancien (1605) d'entre eux et nous avons eu la chance de pouvoir l'admirer pendant la conférence, puisqu'il est conservé à Bruxelles (voir photo ci-dessus). Dans son exposé, S.R. Sarma décrit les différentes parties, et leurs fonctions, d'un astrolabe sanskrit en insistant sur les particularités qui le rendent différent des astrolabes indo-persans, arabes ou européens. Il montre à quel point cet astrolabe est proche des premiers manuels sanskrits (XIVe siècle) sur la construction de l'astrolabe. En particulier le *Yantrarāja* de Mahendra Sūri, un ouvrage très populaire dont il subsiste plus de cent manuscrits et dont S. R. Sarma a donné des extraits. Cette proximité, parmi d'autres arguments, lui permet de confirmer que l'astrolabe conservé à Bruxelles date bien du début du XVIIe siècle, contrairement à l'opinion de David Pingree et de son ancien propriétaire, le Time Museum de Rockford, Illinois.

F. Patte (*Rythmes et algorithmes - Le génie mathématique indien*), en analysant le traité *Saṅgitratnākara* de Śārṅgdeva (XIVe siècle), montre comment les mathématiciens indiens modélisaient la rythmique. Il s'agit d'étudier tous les rythmes possibles de longueur n constitués de sons de longueurs 1, 2, 4 ou 6. Pour ce faire, Śārṅgdeva étend les algorithmes utilisés depuis Piṅgala (3e siècle av. J.C.) à la résolution des problèmes suivants : prastāra (produire la liste complète des rythmes possibles), saṁkhyā (calculer le nombre des rythmes possibles), naṣṭa et uddiṣṭa (construire un rythme connaissant sa position dans le prastāra et inversément), et meru (construire un tableau donnant le nombre de rythmes comportant autant de brèves, etc.). Dans son article, F. Patte illustre à l'aide d'exemples et explicite par des formules l'algorithme de résolution des deux premiers problèmes.

Comme K. Ramasubramanian dans son article astronomique, K. Mahesh, R. V. Pai. et K. Ramasubramanian (*Mādhava series for π and its fast convergent approximations*) s'intéressent aux découvertes des mathématiciens du Kérala. Ainsi, Mādhava (c. 1340-1425) propose différentes séries permettant d'obtenir rapidement plus de décimales exactes de π que n'en permettait la méthode, traditionnelle depuis Archimède, des polygones inscrits et circonscrits. Ces séries préfigurent celle de Gregory-Leibniz pour π et le développement en série de Taylor de la fonction arc tan x. Les raisonnements qui y mènent, rapportés par la *Yuktibhāṣā* de Jyeṣṭhadeva (XVIe siècle), ne reposent pas sur un calcul différentiel déjà maîtrisé, mais mêlent de manière subtile les propriétés des triangles semblables, les sommes de séries entières et quelques convergences bien choisies. Pour terminer, un extrait du grand commentaire *Āryabhaṭīyabhāṣya* (c.1500) de Nīlakaṇṭha atteste très clairement de la conviction qu'a son auteur de l'irrationalité de π.

Enfin, J. M. Delire (*Entre astronomie et mathématiques, les découvertes indiennes en trigonométrie : la construction des tables de sinus et quelques méthodes d'interpolation*) reprend l'histoire du sinus, que nous devons aux mathématiciens indiens, depuis les plus anciens siddhāntas jusqu'aux travaux des mathématiciens du Kérala. Après avoir montré comment construire une table de sinus des arcs multiples de 3°45' à l'aide des propriétés des angles moitiés et complémentaires (*Pauliśasiddhānta*), l'article souligne l'originalité de la table de différences de sinus de l'*Āryabhaṭīya* (499), basée sur les propriétés des différences secondes de sinus. Ces propriétés sont utilisées plus largement par les mathématiciens du Kérala cités ci-dessus, auxquels il faut ajouter Śaṅkara Vāriyar (c.1500-1560), auteur d'un commentaire du *Tantrasaṅgraha* de Nīlakaṇṭha et du *Kriyākramakarī*, un commentaire incomplet de la *Līlāvatī*, complété par Nārāyaṇa (c. 1540-1610). Contrairement à Āryabhaṭa, qui se contente d'énoncer la méthode de construction des différences tabulées de sinus, Śaṅkara et Nārāyaṇa explicitent celle-ci selon un procédé indien que l'on pourrait comparer à une preuve, l'upapatti. Ils font de même pour une méthode d'interpolation du second degré permettant de calculer le sinus d'un angle non tabulé. Malheureusement, bien que présentant les trois premiers termes de la formule de Taylor pour le sinus, cette dernière méthode ne semble pas pouvoir produire les termes suivants.

Il me reste à remercier tous les intervenants de cette journée d'étude, en commençant par les auteurs cités ci-dessus auxquels il faut ajouter Ch. Minkowski (Boden Professor of Sanskrit, University of Oxford), pour son exposé « *Sanskrit Astronomers and the Mughals* », qui nous ont fait l'honneur d'accepter de venir à Bruxelles. Mes remerciements vont aussi au FNRS, à la Fondation Wiener-Anspach de l'ULB, à la Commission CHAMA de l'International Union of History and Philosophy of Science et à l'entreprise de systèmes optiques AMOS, sans l'aide desquels il aurait été impossible d'inviter les orateurs résidant à l'étranger. Merci aussi au propriétaire des instruments astronomiques anciens qui a permis, à titre totalement gracieux, que ceux-ci soient exposés et accessibles durant toute la journée d'étude. Enfin, je veux saluer Mmes A. Tihon, Cl. Bautze-Picron et MM. J. Doyen et Ch. Vielle pour l'aide qu'ils ont apportée dans l'organisation de la journée, ainsi que les étudiants P. Delire, G. Lucena Gómez et Ph. Bauer Raposo pour leur aide logistique. Et – last, but not least – je voudrais spécialement remercier B. Lemaire pour son excellent et long travail de mise en forme des Actes présentés ici.

J. M. Delire,

Centre Altaïr d'histoire des sciences et des techniques (ULB)
Chargé du cours Science et civilisation de l'Inde - Textes sanskrits (IHEB)

The reality of Indian astronomy

Raymond Mercier
Department of History and Philosophy of Science, Cambridge University

Background

Indian astronomy has a complex history, and both the pre-scientific phase, that is before *ca* A.D. 500, and the scientific period, which continues from that time until the 18th century, have generated many problems and disputes. The problems concern both the qualities of the discipline as pursued within India, and the course of transmission from the Greek world to India.

Here I am concerned with the delineation especially of a recent acrimonious dispute between Roger Billard and David Pingree. In order to present this it is necessary to explain, at least in outline, the peculiar characteristics of the Indian discipline.

If I say that the subject as a scientific discipline begins with Āryabhaṭa, who we know was born in 476, that already gets us into disputed territory. For according to Pingree the first canon that defines the scientific phase is to be found elsewhere - but of that, more later.[1]

Āryabhaṭa

Āryabhaṭa is known to have written two works, of which that named Āryabhaṭīya[2] defines the principal canon of his astronomy, that is the set

[1]Another written review of this matter, in some respects more comprehensive than this article, in Mercier (2011).

[2]This is the name that came to be used, but which cannot be the original name.

of parameters for the mean and true longitudes of the Sun, Moon and planets. This work contains much besides by way of geometry and arithmetical procedures, and this provides us with a very strong platform on which to build our understanding of the state of mathematical science at that time in India. He wrote another work as well, but although this does not survive as such, we do know the canon of parameters which was given there, since this is preserved for us in the work of another author, who wrote somewhat later in the 6th century, Varāhamihira.

The parameters of the mean motions, as given in the Āryabhaṭīya are as follows

Āryabhaṭīya: bhagaṇa		
Sun	4320000 t'	t' = t/1577917500 t = days from Kaliyuga Dawn t_{KY} = -3101 Feb 18 Dawn, JD 588465.75
Moon	57753336 t'	
Node	0.5-232226 t'	
Apogee	0.25+488219 t'	
Mercury	17937020 t'	
Venus	7022388 t'	
Mars	2296824 t'	
Jupiter	364224 t'	
Saturn	146564 t'	

Figure 1: The canon of the Āryabhaṭīya

In order to compute any one of these mean longitudes the time t, in days, is found, that is the number of days from t_{KY} = -3101 Feb 18, 6 a.m. This point, dawn of the era Kaliyuga, is the origin point of the chronological system. In more practical terms the canon is based on the epoch 499 Mar 21 Noon. The interval between that moment and t_{KY} is 3600 years, indeed this is *exactly* 3600 years when the year is measured by that given by the canon.

The integers presented in the table, 4320000, etc, are termed *bhagaṇa*, literally 'revolution numbers'. Thus in 1577917500 days, the Sun makes 4320000 revolutions, so that the length of the year is 1577917500/4320000 = 365.258680555...days, better expressed as 6,5;15,31,15, since these calculations are more perspicuous in sexagesimal form.

Mythical chronology

The Kaliyuga to which Āryabhaṭa refers is part of a fourfold scheme familiar from the Purāṇas. There are 18 works of that name, works of what might be called 'folk history', combined with guidance to Hindu worship, originally composed, according to Pargiter, in Prakrit, but eventually Sanskritized.[3] The scheme used by Āryabhaṭa is an acknowledgement of the traditional Indian chronology, found in numerous chapters of the Purāṇas. This chronology proposes a complex structure of time leading up to the Kaliyuga, the final stage of a fourfold arrangement. In the *Matsya Purāṇa* (Ch. 142), for example, we have, beginning with the kṛtayuga[4].

kṛtayuga	1728000	=	8,0,0,0; years
tretayuga	1296000	=	6,0,0,0; years
dvāparayuga	864000	=	4,0,0,0; years
kaliyuga	432000	=	2,0,0,0; years
caturyuga	4320000	=	20,0,0,0; years

The first of these was a sort of Golden Age, free of strife, after which the world degenerated, reaching, after great wars, the present age, when history as we know it begins.

The great Epic *Mahābhārata* makes some references to this chronological scheme, at least in Book III, *Vanaparva*, the Book of the Forest, and elsewhere.[5] However it has recently been argued by Gonzalez-Reimann that in the Epic the four names of the yugas, kṛta, etc., are not used with the meanings as later fixed, and that some of the passages that make such references were later accretions; certainly it is accepted that there were many such accretions.[6] It is interesting, for example, to consider the passage

> With the degeneration of people, the forces that prosper the world decline, and the Laws produced by the decline of the world are perverted into prayers. Thus is described the kaliyuga that will shortly be at hand; those who live long conform to the Eons they live in.[7]

[3]Pargiter (1918), p. X, 'There are clear indications that the Sanskrit account as it exists in the Matsya, Vāyu, and Brahmāṇḍa was originally in Prakrit, or, more accurately, that it is a Sanskritized version of older Prakrit ślokas.'

[4]In later literature this is renamed Satyayuga, 'period of truth'.

[5]Book I, 85; III, 148, 158, 186, 187; V, 130, 140.

[6]Gonzalez-Reimann (2002).

[7]Mahābhārata III, 148; van Buitenen (1975), p.506.

This is a curious self-reference, rather as if we were to say that people in the Middle Ages knew that they were living in the Middle Ages ! There has been a habit of supposing that the Kaliyuga began just when the wars of the *Mahābhārata* were completed, but it is just this rather simple view which Gonzalez-Reimann has contested.

In the Purāṇas, at least, the scheme finds its full expression. Here, however, although we find innumerable genealogies, representing a sort of history or chronicle, it lacks anything we would regard as a secure chronological frame, that is, some expression such as a year or date in relation to the familiar Śaka or Vikrama epochs. Since the Purāṇas come down to roughly the 4th century A.D., by which time there are inscriptions clearly dated by one or the other of those epochs, it is strange that no such dating occurs in any of the Purāṇas.

The many chapters of the *Matsya Purāṇa* include many references to calendars and to the solar system, but always from the point of view of Hindu religious practice, and without any full technical formulation. Nevertheless there are many remarks which indicate contemporary technical practice with the calendar, by which Hindu worship is linked to the New Moon (Āmavasya), and Full Moon (Pūrṇimā), with days counted from Pratipada (first tithi); the nakṣatra are also listed. The five year cycle of the Vedaṅga Jyotiṣa is known, with the names of the five years of the cycle.[8] In that respect these works remind me of the Talmud, which can serve as a guide and compendium to religious life, but where references to the Jewish calendar, for example, are essentially devoid of technical value.

Calendrical references in Āryabhaṭa

There are only four passages in the *Āryabhaṭīya* which serve to fix the treatise in relation to Indian chronology and calendar.

Gītikāpada 3-4. The canon is said to commence at Aries on Wednesday at sunrise at Lāṅka.

Gītikāpada 5. The current *kalpa* begins on Thursday, 6 Manus + 27 yugas + 3 quarter yugas before the beginning of the current Kaliyuga.

[8] *Matsya Purāṇa*, Chs. 54 (nakṣatras), 141 (five year cycle, the course of the month, with details of crescent visibility), 142 (yuga chronology).

Kālakrīyapada 11. The year, month and day begin simultaneously with the light half of Caitra.

Kālakrīyapada 10. Āryabhaṭa is age 23 when 3600 years of current yuga have elapsed.

None of these data however suffice to fix the epoch of the canon securely in terms of any historical epoch, such as Vikrama or Śaka. However the contemporary author Bhāskara I, in his *Mahābhāskarīya* I.4, specifies that 3179 is to be added to the lapsed years of the Śaka date to obtain the years of the Kaliyuga. Since the Śaka era, a true historical epoch, dates from A.D. 78, this fixes the Kaliyuga as 3101 in excess of the A.D. year, so that the Kaliyuga begins in -3101. From the remaining details of the passage *Mahābhāskarīya* I.4 we can determine the precise equivalence

0 KY = -3101 Feb 18 dawn (588465.75).

In terms of the Indian calendar this is day 1 of the bright half of Caitra, as calculated by the true Sun.[9]

Since the year by this canon of Āryabhaṭa is 6,5;15,31,15, then 3600 (=1,0,0;) years is 6,5,15,31;15 = 1314931.25 days. When this interval is added to the epoch of the Kaliyuga we have 588465.75+1314931.25 = 1903397, that is Noon of 499 Mar 21. Since 499 Mar 21 is Sunday, then -3101 Feb 18 is Friday. That is the Kaliyuga is precisely 3600 years earlier than the epoch used by Āryabhaṭa.

The interval of 3600 years between Āryabhaṭa's practical epoch 499 Mar 21 and the precise date of the Kaliyuga shows beyond doubt that *the Kaliyuga was indeed fixed in just this way by Āryabhaṭa*, even though we need to fill in the details from Bhāskara's work. In other words he may be taken as the first to convert this mythical epoch in terms of a real practical calendar. Although Āryabhaṭa has not given in his text, as far as it is extant, this definition of the Kaliyuga, we are perfectly justified in attributing it to him. In this way the Kaliyuga, hitherto only vaguely used in the Purāṇas, is henceforth *by definition 3600 years prior to Āryabhaṭa's chosen epoch in 499*. Ever since the time of Āryabhaṭa the Kaliyuga has been used in this precise way, but unfortunately, this precise identification of the epoch has been so frequently applied also to all earlier uses of the term, and in particular to dating the wars of the *Mahābhārata*.

[9]By 'bright half' is meant the fortnight beginning at the New Moon. The 'dark half' is the remainder of the month, beginning at the Full Moon.

Why would Āryabhaṭa have adopted the interval 3600 years? Although he may have had some foggy notion of how far into the past he might conceive of the 'olden days', we can reasonably speculate that he was influenced by chronological ideas coming from Babylonian chronological schemes. The interval 3600 presumably reflects Babylonian and related Hellenistic speculation. Indeed there was a cuneiform character for this number, ŠÁR 𒊹, that formed a basic unit in the Ante-diluvian King lists in Berossos.[10] The point is that Āryabhaṭa now gives for the Kaliyuga a precise chronological value; for the first time it is tied to historical reckoning, in contrast to whatever vague notions might have been found in the Purāṇas. The very division of time into the four periods, kṛtayuga, etc., is expressed as we have seen in units that are so readily expressed in Babylonian sexagesimal form 8,0,0,0; etc., that we feel compelled to admit Babylonian influence.

On reflection, the simplicity of the relation between the Kaliyuga at -3101 and the epoch of the Āryabhaṭīya is truly compelling, and admits no alternative to the view that this date of the Kaliyuga is simply due to Āryabhaṭa. Gonzalez-Reimann remarked that 'in all likelihood, the astronomers of the Gupta period who gave the yuga system a chronological framework by establishing that the present Kali Yuga had started in the year 3102 B.C.E., a date that was arrived at by calculating when the last conjunction of all the planets at the beginning of the constellation Aries had taken place.'[11] This was supported, however, not by the argument given here, but by a reference to Pingree, who seriously imagined that such a conjunction could have been derived from the *Almagest* for the time of the Kaliyuga.[12] Pingree found from the *Almagest* the mean longitudes Saturn 290;48, Jupiter 325;4, Mars 301;55, Sun 314;38, Moon 323;2. These data, although correctly calculated for -3101 Feb 18 6 a.m., show a spread of 35 degrees, and hardly suggest any sort of general conjunction at that time. However Pingree persuaded himself that

> If one knew from a Greek set of tables, the "correct" mean tropical (or sidereal) longitudes of the planets at some time in the fifth century and their period relations, one could derive for each by means of an indeterminate equation an integer number of rotations in a Kalpa such that the conjunction of -3101 and the "correct" mean longitudes in the

[10] Verbrugghe (2001). p.70.
[11] González-Reimann (2002). pp.169, 189.
[12] Pingree (1976), p. 115; Pingree (1981), p. 555.

fifth century would both be accounted for. This, I suggest, is what the Indians did.

It is difficult to understand how anyone could be taken in by this sublime nonsense. Gonzalez-Reimann showed much critical acumen in his review of the chronological framework of the Epic, only to lapse when faced with Pingree's claim.

The Sūrya Siddhānta

Āryabhaṭa is accepted as the author of another canon, closely related to this one, and which Varāhamihira named Sūrya Siddhānta[13]. One of the distinctions from the canon which we have been discussing is that although the practical epoch is still 499 Mar 21 noon, it is now referred to Kaliyuga at sunset, in contrast the the Kaliyuga at sunrise used above.

Āryabhaṭa, Sūrya Siddhānta: bhagaṇa		
Sun	4320000 t'	t' = t/1577917800
Moon	57753336 t'	t = days from
Node	0.5-232226 t'	Kaliyuga midnight,
Apogee	0.25+488219 t'	t_{KY} = -3101 Feb 17/18, mdnt.,
Mercury	17937000 t'	JD 588465.5
Venus	7022388 t'	
Mars	2296824 t'	
Jupiter	364220 t'	
Saturn	146564 t'	

Figure 2: The canon of the Sūrya Siddhānta

The differences from the previous canon are in the number of days in the mahāyuga, as well as in the bhagaṇa of Mercury and Jupiter. The consequence of the change in the number of days is that the year is now 1577917800/4320000 = 365.25875... days, better expressed as 6,5;15,31,30. This small difference is however revealing, for in 3600 such years there are 6,5,15,31;30 = 1314931.5 days. When this interval is added to the epoch of the Kaliyuga we have 588465.5+1314931.5 = 1903397, that is again Noon of 499 Mar 21. In other words the Kaliyuga, but this time as a midnight epoch,

[13]Sastry (1993), *Pañcasiddhāntikā*, Chs. IX, XVI, XVII; Billard (1971), p. 73 seq.

has again been derived from the true epoch 499 Mar 21 Noon. Nothing could make it more transparent that the Kaliyuga is a purely artificial derivation, being set at dawn or midnight to suit the a priori construction of the canon.

Billard drew attention to the simple relation between the two versions of the Kaliyuga and the lengths of year of these canons, writing 'En fait ce moment 3600 ans KY est une articulation de système des yuga mise à profit pour procéder à un aménagement des constantes, mais qui n'a pas nécessairement une signification chronologique très étroite.'[14]

The general conjunction of Sun, Moon, and planets at the time of the Kaliyuga, which is assumed in both canons is purely hypothetical, set as a purely *a priori* constraint, but which is not supported by any empirically sound system of chronology. Its adoption by Āryabhaṭa illustrates only his obeisance to the received Hindu tradition, so evident in spite of his boldness in other respects.

The Yuga system of Āryabhaṭa

According to Gītikā 5, he defines the period Manuyuga as 72 yugas. This corresponds to the period Manvantara of the puraṇic chronological schemes, although in the *Matsya Purāṇa* (Ch. 142), it is 71 times the mahāyuga of 4320000 years. The Kalpa, the Day of Brahma, in turn, is 14 times the Manuyuga, both by Āryabhaṭa and the Purāṇa. Historical time is supposed to be situated somewhere in the middle of this Day of Brahma, in such a way that the interval from the beginning of the Kalpa to the Kaliyuga is

6 Manus + 27 yugas + 3 quarter yugas = $(6\times72 + 27\ 3/4)$ yugas, which is
$(459\ 3/4)\times4320000 = 1{,}986{,}120{,}000$ years or
$(459\ 3/4)\times1577917500 = 725447570625$ days.

Since $725{,}447{,}570{,}625 \equiv 1 (\mathrm{mod}\ 7)$, it follows that the first day of the Kalpa is Thursday, as he says in Gītikā 5.

In the *Matsya Purāṇa*, the Kalpa is constructed differently:

Manvantara = $71\times4320000 = 306{,}720{,}000$ = 23,40,0,0,0; years.
Kalpa = $14\times$Manvantara = $4{,}294{,}080{,}000$ = 5,31,20,0,0,0; years.

Compared with the astronomical canon of Āryabhaṭa the singular difference is that in the Purāṇa there is no mention of *historical* time, no way in which a historical epoch, or clearly recognizable historical event, is placed in this grand cosmic scheme. This is truly typical of the grand chronologies

[14]Billard (1971), p. 80.

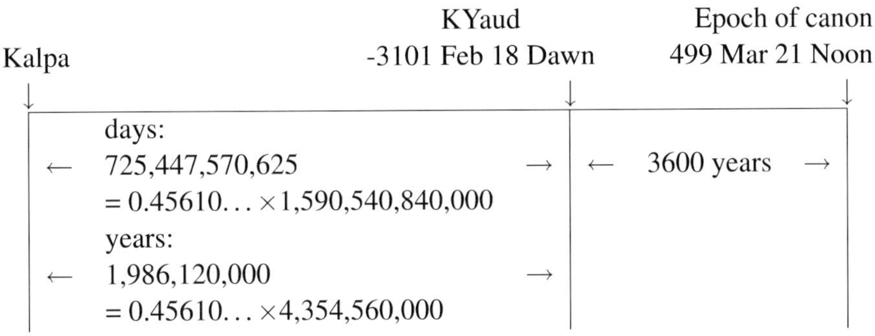

Figure 3: Chronological frame of *Āryabhaṭīya*. KYaud = Kaliyuga audāyika, Sunrise Kaliyuga.

of the Purāṇas, and we will see another most important example later in the *Viṣṇudharmottarapurāṇa*, a work that was important in Pingree's supposed historical reconstruction of scientific Indian astronomy.

There is a puzzle, perhaps, in the statement in Gītikā 3-4, where after stating the bhagaṇa for the Sun, Moon and planets, it is stated that the revolutions started in Aries on Wednesday at sunrise. For we have just noted that the first day of the Kalpa is Thursday, since 725,447,570,625 $\equiv$ 1(mod 7). For if the mean Sun is zero at dawn at the Kaliyuga, it was zero also at dawn at a date which is earlier by an integral number of years. It must be the case therefore that he has in mind the *true* Sun on that day, as is indeed remarked by the later commentators, such as Parameśvara.[15]

It is clear however that these a priori speculations about the Manvantara and the Kalpa have no bearing on the astronomical reality of the canon, since the conjunction at the Kaliyuga is the lynch pin of the system.

Inscriptions

Inscriptions in India are dated in a wide variety of ways, yet if not in a regnal year or a local era, they are more likely than not to be dated by either the Vikrama or Śaka eras. Occasionally however one finds one dated by the Kaliyuga. Here is one example of that, the Aihole inscription of the Chalukya King Pulakeśin II (AD 610-642); Aihole is in the northwest Deccan. The year is specified both as Kali 3735 and as Śaka 556 (AD 634-5).

triṃśatsu tri-sahasreṣu bhāratādāhavāditaḥ

[15] Kern (1874), commentary to gītikāpada 4 [but verse 2 in Kern's edition, p. 7].

saptābda-śata-yukteṣu śa[ga]teṣu pañcasu ||33||
pañcāśatsu kalau kāle ṣaṭsu pañcaśatāsu ca
samāsu samatītāsu śakānāmapi bhūbhujām ||34||

33. (*Now*) when thirty (*and*) three thousand and five years besides, joined with seven hundred years, have passed since the Bharata war;

34. And when fifty (*and*) six and five hundred years of the Śaka kings also have gone by in the Kali age;

This long inscription of some 128 lines is of immense historical interest quite apart from its unusual use of the Kaliyuga for the year. A full translation was given by Fleet.[16]

Although the use of this Era is unusual Fleet has noted six others where it is used in the South Indian group from various periods up to the twelfth century, and a few others from the Northern group.[17]

There is a number of exceptional collections of dated inscriptions, all of which are of course of great interest in confirming the applications of the astronomical canons. For inscriptions dated in the Śaka and Vikrama eras we have a series of articles by Kielhorn in the *Indian Antiquary* in 1890-1 (Vikrama), and 1894-7 (Śaka).[18] There are further collections by Kielhorn in Appendices to *Epigraphia Indica*, volumes V and VII.[19] There is finally a much more recent collection by Bhandarkar (1983).[20]

Inscriptions in South East Asia as well were dated by the Indian canons, or canons very closely related to them, and it is interesting to recall that Roger Billard's studies began with his work on those at Phnom Penh in Cambodia.

Deviation curves

The work of Roger Billard has many singular features betraying a most attentive and intelligent reading of a complex literature.[21] At the heart of his work, the feature that makes it so important, is a simple, even inevitable, development. It is a commonplace to test an ancient astronomical work by a recalculation from modern parameters of the state of the sky at that time,

---

[16]Sircar (1983), 'The Chālukyas of Bādāmi', p. 443-450; Fleet (1876).
[17]Fleet (1911).
[18]Kielhorn (1891), Kielhorn (1897)
[19]Kielhorn (1899), Kielhorn (1903)
[20]Bhandarkar (1983).
[21]Billard (1971).

indeed how else would one hope to date an ancient work? Such a procedure is absolutely banal, and numerous examples of it are to be found in the papers of the late Otto Neugebauer, such as his dating of numerous Greek horoscopes.

In Billard's study of the Indian canons he created a systematic application by considering the difference quantity 'ancient mean longitude minus modern mean longitude', or briefly the *deviation* of the mean longitude. The mean in question may be that of the Sun, Moon, or any of the planets, and also of course the lunar node and lunar apogee. Further one can as well consider the deviation between ancient and modern *true* longitudes. The *deviation curve* is simply the graph of the deviation against time.

Billard's first application was indeed to the *Āryabhaṭīya*.[22] The mean longitudes are taken from the formulae given above, and the resulting deviation curves are shown in Figure 4. The singular feature which immediately strikes one is that the bundle of curves (with the exception of Mercury) form a very tight node at one time, very near the year 500. That is to say, the agreement of all the mean longitudes (except that of Mercury) together at this date means that the predictions from the canon were accurate at only one time, where the curves form the node, and therefore that the observations on which the canon was based must have been conducted around that time, and successfully reduced to the reference date 499 Mar 21. The curves shown here have been produced by a program of mine which makes use of programming techniques for Windows, which can be carried out perfectly easily on a modern desktop PC, in contrast to Billard's work which was carried out in the 1960's on a main frame in Paris, to which he obtained access.

In order to understand better what is represented by these curves we note first of all that the mean longitudes of the Indian canon are sidereal longitudes, measured from some point in the ecliptic fixed in relation to the stars. Such an origin point is completely arbitrary, it must be emphasized. Secondly the modern mean longitudes are tropical, that is measured from the equinox of date. It is immediately clear that the curves will have a slope that is certainly of the order of magnitude of the rate of precession, roughly one degree in 70 years. If the Indian longitudes were as accurate as the modern then they would all have the same slope and would coincide. It is a special feature of the present situation that this is certainly not the case, for we can see the well defined node. If we note further that the mean longitudes of the canon are all zero at the time of the Kaliyuga (apart from the lunar node,

[22] Billard (1956).

which is 180°) we can appreciate that this imposes an arbitrary a priori constraint on the curves, for to be sure the correct values at that time were hardly in agreement with such a drastic assumption. Since the curves nevertheless come together at least at the node it is clear that a further constraint has been imposed at the time of the node, a constraint due simply to control by genuine observations. That was Billard's conclusion, and it is incontestable. Nevertheless Pingree was unable to appreciate this, partly due to a lack of scientific intelligence, partly because it would have forced him to abandon certain cherished views which he had developed shortly before Billard's discoveries. Pingree's arguments will be outlined in another section.

These modern means were computed on the assumption that they are for the central meridian in India, taking the meridian through Ujjain in Madhya Pradesh, which was generally reported in the Indian texts as having been used for this purpose. We extend this calculation of deviation curves to include the dependence on the longitude of the reference meridian, so that we may find perhaps an optimum meridian just as the curves shown in Figure 4 point to an optimum year. If the meridian of one of the modern means is changed then the curve is displace up or down by the amount (daily mean motion)×(change of meridian). Thus for the Moon a change of 10 degrees in the meridian causes a shift of about 0;20, while for the Sun the shift is only some 0;2, and for the planets it is even less. Therefore there will be an optimum meridian which causes the node to be at its tightest.

Let $D_n(t, \phi)$ be the deviation of the nth mean at time t and meridian ϕ; n runs from 1 to 9. Then there are two ways to measure the tightness of the node. In the first we simply measure the deviations from the x-axis, disregarding the fact that we are measuring sidereal against tropical means. In the second we consider the node as an intrinsec feature within the curves so that we compare D_n against the mean of the deviations $\overline{D} = (\sum_I D_n)/N$, where I is the set of deviations to be considered, and N is their total number. In these two approaches we have measures Q_1 and Q_0, essentially mean square measures of the tightness of the node.

$$Q_1 = \sum_I D_n^2 \qquad Q_0 = \sum_I (D_n - \overline{D})^2$$

Billard introduced a simple notation to indicate the subset I, and whether Q is absolute or relative. For example (0 1111 00000) means that only the solar and the three lunar deviations are selected, while the first 0 indicates that Q_0 is calculated.

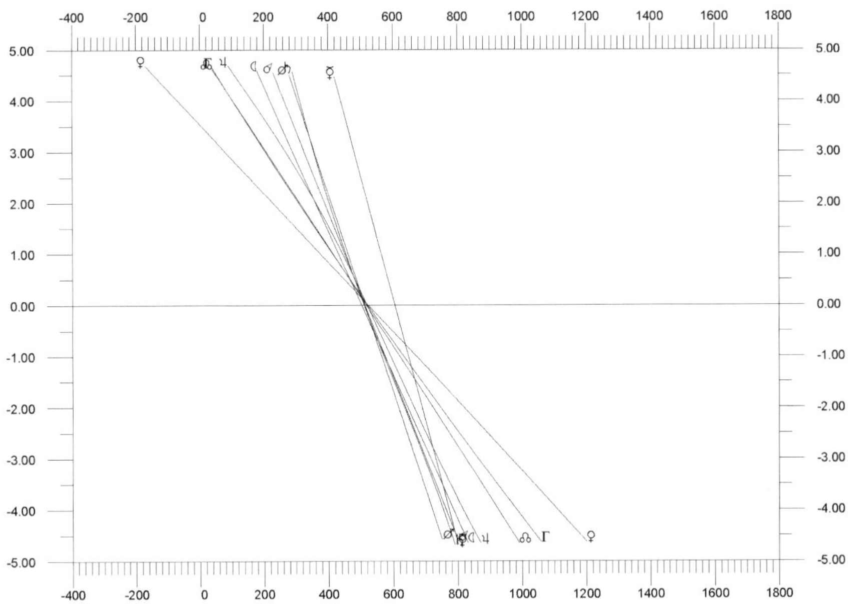

Figure 4: Deviation curves of the Āryabhaṭīya

Legend:						
	☉	Sun	☊	Lunar node	♂	Mars
	☾	Moon	☿	**Mercury**	♃	Jupiter
	Γ	Lunar apogee	♀	Venus	♄	Saturn

The quantities Q_0, Q_1 depend on the year t and the meridian ϕ. When the node is tightest the mean square deviations is minimum. Near this minimum point, Q as a function of t and ϕ is represented approximately by the quadratic expression,

$$Q = h_{11}(t - t_0)^2 + 2h_{12}(t - t_0)(\phi - \phi_0) + h_{22}(\phi - \phi_0)^2 + h_{00}.$$

The statistical properties of the bundle can be computed from the coefficients h_{nm}. There is a detailed survey of results for a number of Indian canons by Mercier.[23] The essential result is that the optimum meridian lies near 80° East of Greenwich, so passing through central India, so that we have a proof that not only were the mean longitudes determined during the lifetime of Āryabhaṭa, but that they were fixed by observations made in India.

Since the function Q is represented in the neighbourhood of the minimum value h_{00} by this quadratic expression it may be envisioned as a surface

[23] Mercier (1987).

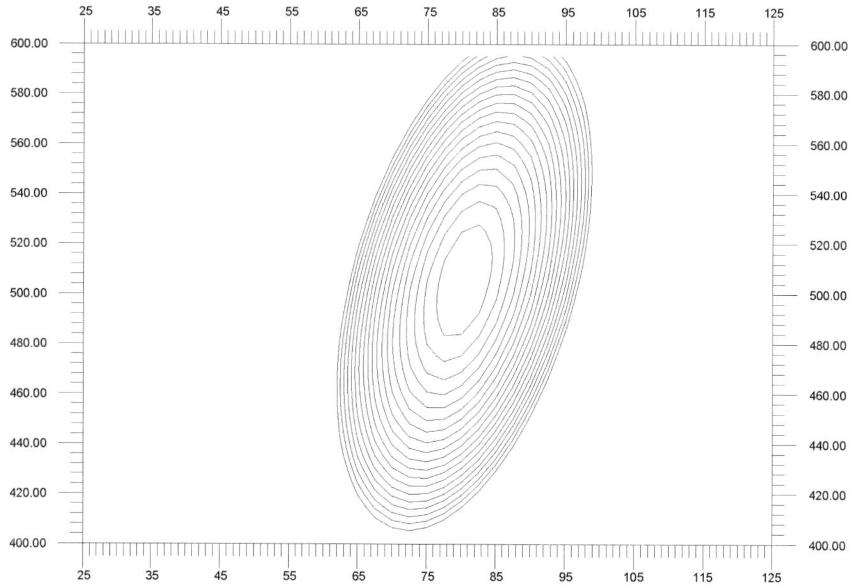

Figure 5: Diagram showing the level curves of Q, calculated for the Sun and lunar deviations of the Āryabhaṭīya. The centre is at year = 498.1±31.6, meridian = 80.8±5.9.

which is approximately a paraboloid near the minimum, and whose cross-sections are ellipses. These ellipses are level curves of the surface, and by routine numerical procedures these level curves may be determined given the values of Q_0. This requires that a good number of values of $D_n(t, \phi)$ are found over a range of t and ϕ, and that an algorithm be employed which derives the coordinates of the level curves. The algorithm employed here is one published by the ACM.[24]

The centre of the contours marks the optimum year and meridian, and the plot in Figure 5, this is the year 498.1±31.6, meridian = 80.8±5.9. The statistical bounds are determined by an application of the Student distribution. Of course the statistics are not Gaussian since the number of elements is at most 9.[25]

[24]http://www.netlib.org/toms/, program 531, "contours through equal values of a surface" by W.V. Snyder, ACM TOMS 4 (1978) 290-294. This code is in Fortran, but converted to C for incorporation into my own program.

[25]Billard erroneously assumed Gaussian statistics in his analysis of such results.

Precision of the parameters

The revolution numbers, the bhagaṇa, are given as integers. How exact are these integer values, that is, how much does the deviation change when a revolution is changed by one?
The interval of 3600 years from the Kaliyuga to 499 is 1/1200 of the mahāyuga:

3600/4320000 = 1/1200.

Saturn, for example, makes 146564 revolutions in 4320000 years, so that in 3600 years there are

146564/1200 revolutions, or

146546×360/1200 degrees = 146546×0.3 degrees.

Thus if the revolution number is changed by one, the deviation in the year 499 is changed by 0.3. This is evidently true of all the mean longitudes. Since even a displacement of 0.3 would destroy the nice coherence of the curves, it is evident that all the revolution numbers have been adjusted to be at their best values, subject only to the *a priori* assumption of the grand conjunction at the Kaliyuga : *The bhagaṇa of Āryabhaṭa are the best integer values possible.*

Observation to fix the mean longitudes

It is evident of course that one does not observe mean longitudes directly. Further, although the mean longitudes are sidereal, that does not mean that one has to observe the planets against the stars, but instead one has only to log the planet against the Sun, and since that is given as a sidereal longitude the planet inherits a sidereal longitude.

The canon of the Āryabhaṭīya includes planetary equations, and these are, it is true, less well adjusted to achieve empirical exactness than was the case with the *Almagest*, for example. This aspect of the canon can be illustrated by the calculation of the deviations of the true longitudes : 'ancient true longitude minus modern true longitude'. For Saturn such a true deviation is shown in Figure 6, where the true deviation of the Sun is also plotted. While the difference from the Sun can reach three degrees in either direction, it is evident that a good balance has been achieved, which is the result of a careful adjustment of the mean planet. In order to achieve such a good result it was necessary to observe the planet, and then make calculations using trial values of the mean longitudes of the planet, until it was clear that on the average a good balance was achieved. This would have meant continuing until perhaps the year 540. Of course there is no information about the date when Āryabhaṭa declared his work to be finished, and he might have continued even until the year 550 before putting his pen down.

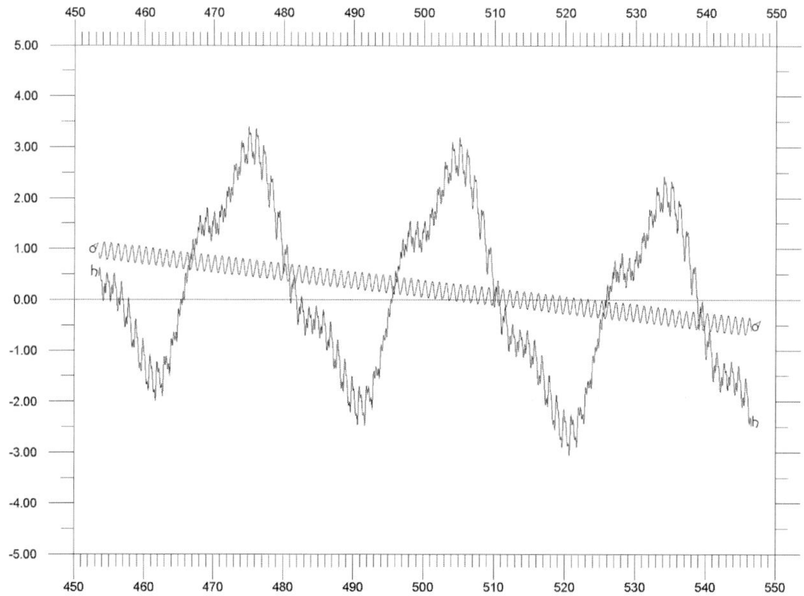

Figure 6: True deviations of Saturn and the Sun, according to the *Āryabhaṭīya*, showing the balance of Saturn around the Sun.

In the case of the *Sūrya Siddhānta*, we can see easily that the bhagaṇa for Jupiter led to a less satisfactory results, and this is evident in the true deviations shown in Figure 7. This weakness was overcome in the *Āryabhaṭīya*, which is quite evidently the later work. Another feature of interest is that the true deviation of the Sun is also much improved in the later work. This is the result simply of a change in the solar apogee from 80 to 78 degrees. In fact the true deviation of the Sun in the *Āryabhaṭīya* is even smaller than with the *Almagest*, in spite of the greater sophistication of Ptolemy's model of the Sun.

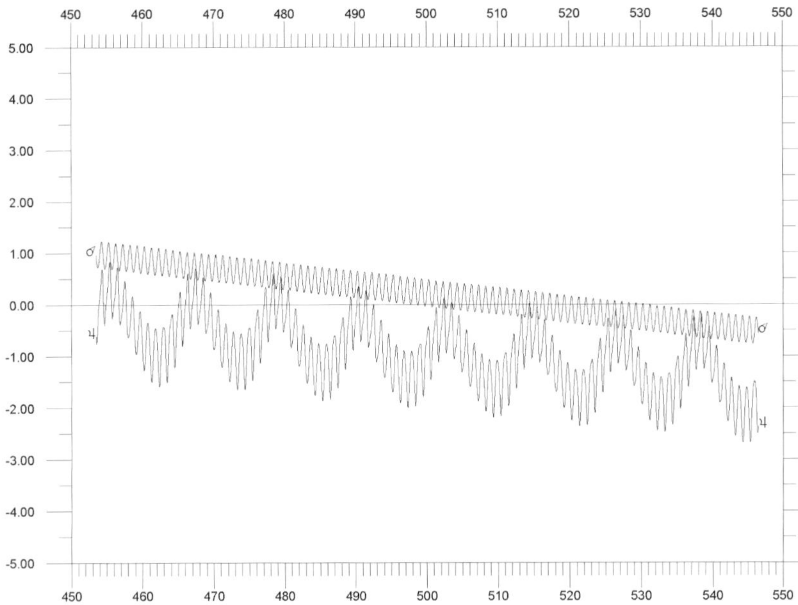

Figure 7: True deviations of Jupiter and the Sun, according to the *Sūrya Siddhānta*, showing the imbalance between Jupiter and the Sun.

Pingree on the Use of 'Greek Tables'

In his article "The recovery of early Greek astronomy from India", Pingree presented arguments intended to demonstrate that Āryabhaṭa could have derived his mean and true parameters from Greek tables extant in his time, ca CE 500.[26] He wrote

> But if Āryabhaṭa did not observe, how did he arrive at mean motions that produce results more correct for his time than any other? The answer is extremely simple: he used Greek tables of mean motions to compute the mean longitudes for a specific time, and thence derive the rotations in a mahāyuga.

He presented a few calculations from the *Almagest* and from the Indian canons, claiming indeed that he was giving the mean longitudes for the planets and the Moon for the epoch date 499 Mar 21 Noon. The numbers which

[26]Pingree (1976), p. 115.

he gave are in the following table, but I have changed the labels of the rows. In his original he used ‘Saturn’, etc. as the labels as if he were presenting the mean longitudes of the planets, and so wanted this table to illustrate that calculations from the *Almagest* gave results much the same as those from the Indian canons. However an easy re-examination shows that he was actually giving the synodic differences.

Ptolemy: *Almagest*		*Āryabhaṭīya*	*Sūrya Siddhānta*
Saturn-Sun	48;40	49;12	49;12
Jupiter-Sun	188;6	187;12	186
Mars-Sun	7;8	7;12	7;12
Venus-Sun	356;45	356;24	356;24
Mercury-Sun	184	186;0	180
Moon-Sun+1	283;30	280;48	280;48
Node-Sun+1	-7;11	-7;48	-7;48
Sun	-2;44,10	0;0	0;0

Figure 8: Corrected version of Pingree (1976): Table 2. The numerical entries are from Pingree’s tables, although the labels have been corrected, as explained here, together with the addition of the last row.

In the first column entitled ‘Ptolemy’ he has in fact listed the quantities shown here in the present Figure 8, now correctly labelled, that is, the *synodic differences* Saturn-Sun etc. Note that there are also errors of one degree in the lunar means, that is, he has presented the quantities Moon-Sun+1, Node-Sun+1. Since at this moment (499 Mar 21) the Sun is zero in the Indian canons it makes no difference whether we call the entries here Saturn or ‘Saturn-Sun’, etc.

Setting aside the possibility of deliberate deception, this is at least confused and confusing. While presenting these quantities as means, he has actually displayed the synodic differences, Saturn-Sun, etc. He wanted the reader to believe that the mean quantities according to the *Almagest* were so close to those of the Indian texts that one could accept his belief that the Indian means were copied from the *Almagest*, or some ‘similar’ Greek source. However since he has displayed synodic differences we have to ask, does the close approximation of the synodic differences actually prove anything? The answer clearly is no, for any two empirically acceptable systems will *inevitably* have the same synodic differences, *so that a similarity demonstrates nothing about any possible historical kinship.*

Pingree omits from the discussion the mean Sun itself, and that is a crucial omission. Here we have a difference of 2;44, which measures the inescapable error between any likely Greek means and the Indian means at this date in 499. In the end it is clear that Pingree's argument is worthless. It is no better than his argument a few pages previously, where he attempts to find a grand conjunction in the year -3101, justified by a calculation from the *Almagest*, which I have noted above.

There is moreover, as we have observed, such a tight agreement within the node of the deviation curves of the *Āryabhaṭīya* that it could never have been achieved except by a canon that has been confirmed by observation at the time in question. Apart from that we have seen that the optimum meridian for the Indian canon is in central India.

Brāhmasphuṭasiddhānta

Brahmagupta (*fl.* 6-7th century) was the author of two influential canons, the *Khaṇḍakhādyaka* and the *Brāhmasphuṭasiddhānta*. The former depends essentially on the *Sūrya Siddhānta* whereas the *Brāhmasphuṭasiddhānta*, composed later, is evidently Brahmagupta's own work. The deviation curves for this show that while there is improvement in so far as the various curves agree much better in their slope, which is because the work is on a broader time base, yet do not so clearly point to a unique observation. Nevertheless they clearly belong to Brahmagupta's period, mid- to late sixth century.

Every Indian canon has a chronological frame, and this is indeed one of the more striking characteristics of Indian astronomy. While the origin of Āryabhaṭa's frame is set at the Kaliyuga, and indeed this mythical prehistoric event is given for the first time a precise sense with his canon, Brahmagupta set the origin very much earlier, so that the base line is 4,320,000,000 years (the so-called Kalpa), instead of the mahāyuga 4,320,000 years of Āryabhaṭa. Brahmagupta's chronological scheme is indicated in outline in the following diagram. Here KYaud means Kaliyuga audayika (sunrise), which is the origin of Āryabhaṭa's frame.

The lengthened time base has the consequence, as remarked, of allowing deviation curves that are more nearly parallel, starting from zero longitude at a far earlier date, but also with the much greater revolution numbers a finer adjustment should be possible, compared with the system of Āryabhaṭa. In this case a change of ± 1 in any revolution number leads to a change of merely $\pm 0.0003°$ in the deviation. However it is evident that this was not

Brāhmasphuṭasiddhānta zero ↓	KYaud (-3101 Feb 18 Dawn) ↓	Śaka (78 Mar 16) ↓
days: ← 720,634,442,715 → = $0.4567 \times 1{,}577{,}916{,}450{,}000$	← 3179 years →	
years: ← 1,972,944,000 → = $0.4567 \times 4{,}320{,}000{,}000$		

Figure 9: Chronological frame of *Brāhmasphuṭasiddhānta.*

exploited to the full by fine adjustments in the bhagaṇa.

This canon was one of those that came to be transmitted to the Islamic world, where it appears as the Zīj of al-Khwarīzmī, early 9th century. Nor does Indian influence in the Arabic sphere end there, for the use of 'Indian' sidereal coordinates continues with the Toledan tables of al-Zarqālī, where the Indian model of precession, ayanāṃśa, becomes 'accession and recession'. This is discussed in the last section of this paper. Beyond that the Kaliyuga even figures in the chronological basis of the Latin Alfonsine Tables ca 1310.

The *Viṣṇudharmottara Purāṇa*

It is a curious fact that the parameters of mean motion of the *Brāhmasphuṭasiddhānta* are listed also in one section of this Purāṇa, a section given the name *Pitāmahasiddhānta*.[27] Pingree regarded this as the *source*, no less, of the mean parameters of the *Brāhmasphuṭasiddhānta*, and this assumption forms one of the major fixed points in his construction of the narrative of the origins of Indian astronomy. As a Purāṇa, it belongs to the class of works known as Upapurāṇas, and is taken by Renou and Filliozat, for example, to belong to the period VII - X century, and 'probablement dépendant de Brahmagupta'.[28] Pingree would have us believe that the *Pitāmahasiddhānta* belongs to the early 5th century, and that Brahmagupta had lifted his parameters from it. A simple inspection of the deviation curves

[27] This Purāṇa has been translated in whole or in part on a number of occasions. See Kramrisch (1928), Pingree (1968), Shah (2002).

[28] Renou, (1947), Tome premier, art. 843.

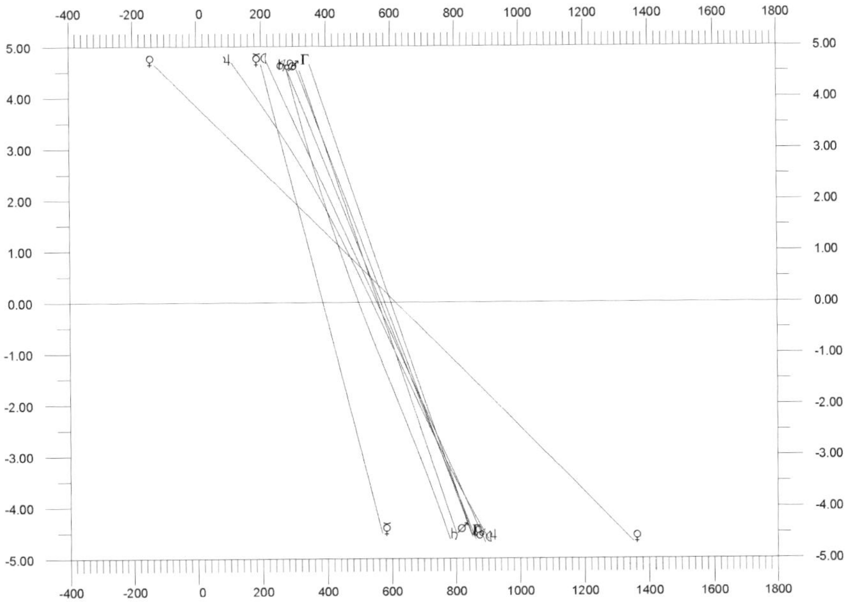

Figure 10: Deviation curves of the *Brāhmasphuṭasiddhānta*.

(Figure 10) for the *Brāhmasphuṭasiddhānta* shows that the parameters belong to the time of Brahmagupta, however.

The quotation of the parameters in the *Pitāmahasiddhānta* is incomplete in two respects. First of all there is no link to historical time, so that these mean parameters are treated just as a sophisticated version of the four-fold yuga construction that is found throughout the earlier Purāṇas, where there is no link whatever to real historical time or events. The second difficulty is that no part of the apparatus of true longitudes is included, that is, nothing of the equations of Sun, Moon, or planets.

As to the first point:

In the *Brāhmasphuṭasiddhānta* the Era, the zero point of the Kalpa, is set at 720,634,442,715 days before Kaliyuga, the same Kaliyuga as defined by Āryabhaṭa. This essential element is specified in *Brāhmasphuṭasiddhānta* I, 26-7. This prescribes a combination of manvantaras, mahāyugas and yugas, amounting to 0.4567 of the Kalpa. In effect:

from the zero of the Kalpa to the Kaliyuga

$$6M+7g+27Y+4y+3y+2y = 1972944000 \text{ years have passed}$$

where

y = 432000 (yuga)
g = 4y = 1728000 (saṃdhi, twilight)
Y = 10y = 4320000 (Mahāyuga, Caturyuga)
M = 71Y = 306720000 (Manu, Patriarchate)
Elapsed: 6M+7g+27Y+4y+3y+2y = 1972944000 years

The solar year in this canon is 1577916450000/4320000000 = 365.2584375 = 6,5;15,30,22,30, so that 365.2584375$\times$1972944000 = 720,634,442,715 days before Kaliyuga.

In the *Pitāmahasiddhānta*, however, there is no such passage, so that the Era is unspecified. That means, of course, that *this text as incorporated in the Purāṇa could never have been used for calculations even for the mean longitudes, never mind the true longitudes.*

Pingree's proposals were published by 1968, just before the appearance of Billard's book, where the analysis of the *Brāhmasphuṭasiddhānta* showed that the parameters of this work belong clearly to the time of Brahmagupta, and so fatally undermined Pingree's 'discovery'. It became evident that Pingree was never willing to withdraw, but went on to insist that the Indians really never founded their work on observation, in spite of the convincing evidence displayed by Billard.

In her recent volume on Indian mathematics Plofker shows herself to be an uncritical devotee of Pingree, a true acolyte, and in particular, regards this *Pitāmahasiddhānta* as 'the first major exemplar of a standard *siddhānta*'.[29] In her discussions of Billard's discoveries she displays what I can only see as a fundamental unwillingness to apply scientific arguments in the evaluation of Indian astronomy.

The Sidereal ecliptic

Tropical longitudes are measured from a well defined point in the ecliptic, the equinoctial point. In contrast, sidereal longitudes are measured from some point of the ecliptic fixed in relation to the stars – but where exactly? In this section we discuss the way this happens in Indian practice. The origin point is clear from the time of Āryabhaṭa, and once it was fixed it was maintained throughout the history of Indian astronomy

The deviation curves of the *Āryabhaṭīya* reveal a well-defined node, which we have discussed. It was clear in Figure 4 that the node lies near the hori-

[29] Plofker (2009), p. 67.

zontal axis. In other words, at that point the sidereal and tropical longitudes are nearly equal. Is that an accident? Yes, it is, even though this proximity suggests the possibility that Āryabhaṭa had defined a new sidereal longitude such that sidereal and tropical coincided at the reference date of his canon, 499 Mar 21. However the Indian sidereal ecliptic was already established and employed in the coordinates of the *nakṣatra*.

It is important to consider carefully the junction stars of the *nakṣatra*, which are found in listings in various Sanskrit treatises, although not in the work of Āryabhaṭa himself. Here we have the positions of the 28 junction stars of the Indian constellations. Their coordinates are, as is well known, the so-called polar longitude and polar latitude. This polar longitude of a star is the point of the ecliptic that culminates simultaneously with the star, while the polar latitude is the arc of the meridian from the star to the ecliptic. The polar longitude is already very much in evidence in the commentary by Hipparchus on the *Phaenomena* of Aratus and Eudoxos, where he gives the polar longitudes of numerous stars that culminate simultaneously with the risings and settings of constellations, always considered in terms of the horizon at the latitude of 36°. In that work Hipparchus considered some 400 stars.

Here I must refer to an important feature of the Greek star catalogue as preserved by Ptolemy, a feature which emerged in an earlier study of mine. This is an exact alignment through the stars α, β Ari, α Vir, and ζ Psc, such that this great circle is also the horizon for the latitude of 36°.[30] The star ζ Psc is within a few arc-minutes of the junction of the ecliptic and the horizon. To be precise, the junction of the ecliptic with the alignment is at -9;23±0;5 on the Hipparchian scale, whereas the Hipparchian polar longitude of ζ Psc is 350;24 = -9;36. Thus between the star and that junction is the very small interval 0;15±0;5.[31] The alignment is shown in Figure 11 where the equinoctial points of Hipparchus and of Indian astronomy are marked ♈H, ♈A respectively, the distance between them being 9;23±0;5. I understand this alignment as a central structural feature of Hipparchian star mapping, a primary coordinate axis. I am at a loss, however, to discover a second such coordinate axis.

The longitudes of the junction stars of the *nakṣatra* define in effect the Indian sidereal longitude, and since the polar coordinates ζ Psc as a junction

[30] Mercier (1977), p. 34. This alignment is very precise for the Ptolemaic coordinates, less so the true positions.

[31] The Hipparchian scale is defined as the Ptolemaic reduced by 2;40.

star are given as 0 that star in effect marks the origin. Actually the precise coordinates of the various stars vary by small amounts from one text to another, so that the polar longitude of ζ Psc is 359;50 according to the listings in the late *Sūrya Siddhānta* and *Śiṣyadhīvṛddhida Tantra*, but 0 in other listings.

In any case we can relate the *nakṣatra* listings to this alignment in the Greek star catalogue : the Indian zero of sidereal longitude is the junction of the ecliptic with the horizon at latitude 36°. Thus this origin point rises with the stars α, β Ari and ζ Psc simultaneously with the setting of α Vir.

The difficulties of a clear confirmation of the coordinates of all the junction stars have long been recognised. We find, however, that if the polar longitude and latitude are computed from the Ptolemaic stars, but with their Hipparchian longitudes increased by 9;23, then there is a remarkable extent of agreement with the listings in the Sanskrit texts, not only for ζ Psc but for many other stars.[32]

The position maintained here therefore is that the sidereal longitude scale employed by Āryabhaṭa is that found in the listings of the *nakṣatra*, and that they in turn are derived from the alignment in the Greek star catalogue. This is not to say that the Ptolemaic catalogue as such was necessarily known in India, but rather that some influence did pass into India from some form of Greek usage, such as is illustrated by the commentary of Hipparchus. The very fact that some of the listings of ζ Psc differ from others by 0;10 is a strong argument in support of an awareness of the details and of alignment, according to which that star is distant by 0;15±0;5 from the origin point.

In a system based on sidereal longitudes precession plays an indispensable role, for without the conversion to tropical coordinates the position of the celestial equator is unknown, and without that all practical observations are impossible. This was well understood by Indian astronomers. Precession is referred to as *ayanacalana* ('motion of the solstice'), and the degree of the motion is the *ayanāṃśa* ('degree of solstice').[33] At the date of Āryabhaṭa the sidereal and tropical longitudes very nearly coincide, as remarked already, and later commentators do remark that for Āryabhaṭa precession is absent.

---

[32] I note in passing that the junction star of Śatabhiṣaj, correctly identified as λ Aqr, is given the negative latitude -0;18, close to the value in reality. According to Ptolemy the latitude is +0;10. This is a curious instance where an Indian record, in spite of its manifest problems, offers a correction to what is clearly a fault in the textual tradition of the *Almagest*; Ptolemy must have had originally the correct figure -0;10.

[33] Perhaps a teminology that refers to the solstice rather than the equinox owes something to the terminology of the familiar section Περὶ τροπῆς in Theon's *Short Commentary* on the Handy Tables, where the Greek astrologers' method of precession is defined.

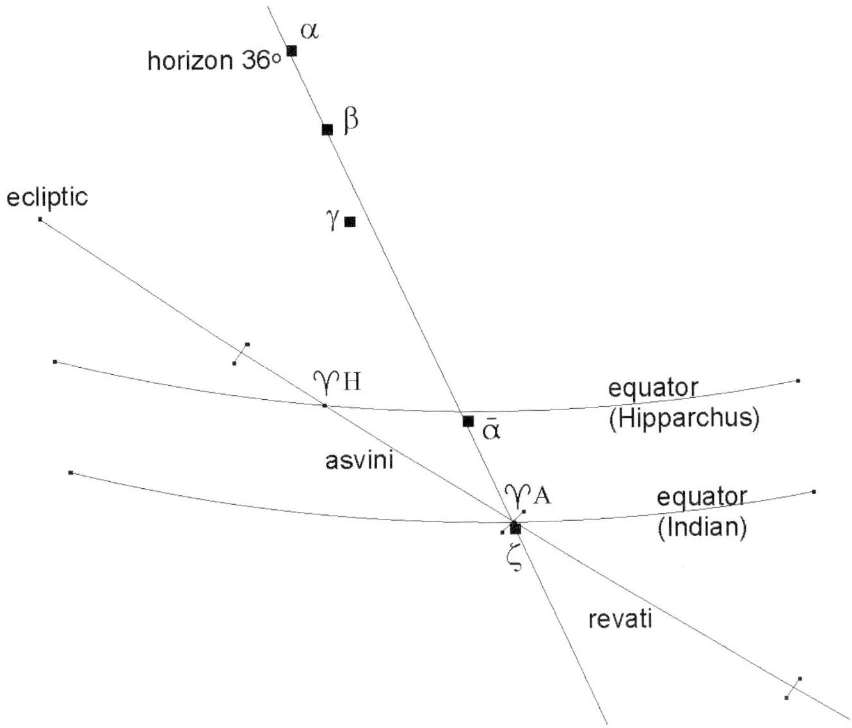

Figure 11: Diagram showing the ecliptic and the positions of the equator at the times of Hipparchus and Āryabhaṭa, along with the alignment through α β Ari, α Vir, ζ Psc, which is also the horizon at 36°, and cuts the ecliptic at the Indian origin of sidereal longitudes near ζ Psc. The point shown as $\overline{\alpha}$ near the Hipparchian equator is diametrically opposite to α Vir.

There is an Indian model of precession, that defined in the later Sūrya Siddhānta, which vanishes just at that date, but indeed all the many Indian models of precession vanish at some point in the sixth century. In the same way in all the canons the deviation curve for the Sun vanishes in that century : this is one of the 'constants' of Indian astronomy, that is, the sidereal Sun is essentially maintained while the Moon and planets are adjusted to suit observations.

If we take the date of Hipparchus as -127, then the time interval to A.D. 499,626 years, will cause a change of ecliptic longitude 9;23 if the rate of precession is 9;23/626 = 53.96" p.a.; this is equivalent to 1 degree in about 66.7 years. This is indeed precession as known from Later Sūrya Siddhānta, where it is presented as a motion from -27° to 27° over 3600 years. This rate of precession is adopted by al-Battānī, who may have learned it from Indian sources. The diagram of Figure 12 shows precession according to the later Sūrya Siddhānta. The Hipparchian origin at year -127, precession -9;23, is marked at a point that lies precisely on the curve of the *Sūrya Siddhānta*.

The passage of the *Āryabhaṭīya* (kālakrīyapada 10), where Āryabhaṭa refers to the epoch date of his canon, is the occasion where many of the commentators draw attention to the question of precession. Nīlakaṇṭha (*ca* 1500), for example, dwells at length on this matter, and records that one '*maṇintha*', surely a Greek Manetho, knew the same model as in the later Sūrya Siddhānta.[34] In any case we can be sure that Āryabhaṭa knew of the phenomenon, since after all the Romaka scholars knew it, according to Bhāskara I (early seventh century), who noted that in his commentary on Āryabhaṭa.[35] The model employed by the later *Sūrya Siddhānta* might well have been known to him.

There are many straws in the wind, yet many problems remain. In particular it would be encouraging if evidence could be found of some use of the alignment with Greek sources. In Greek we have only the astrologer's precession as reported by Theon. This is a zigzag model, and roughly similar to the Indian models, but if it refers to any star as the origin point that would appear to be Regulus, α Leo. Not only is the model applied to the *Handy Tables*, in which the stars are measured from that star, but the origin point in time is -158 when the longitude of Regulus is very near to 120, at least according to a correct modern calculation.

Accession and recession in Arabic sources

The Indian convention of sidereal longitude was adopted to some extent by Arabic astronomers, notably al-Zarqālī of Toledo and Cordoba in the late eleventh century. His work appears in two tracts, which survive respectively in Latin and Hebrew, although both were of course composed in Arabic. The work which survives in Hebrew, and which has been translated into Spanish by Millas-Vallicrosa, might have been entitled something like Tract on the Movement of the Fixed Stars, but in fact the Hebrew text carries no

---

[34]Nīlakaṇṭha, commentary to *Āryabhaṭīya*, kālakrīyapada 10; Pillai (1957), vol. 2, p.13. Varāhamihira, in his *Bṛhat Jātaka*, VII, 1, mentions Maṇittha, Maya, Yavana and Paraśara.

[35]Shukla (1976), Part 2, pp. lviii-ix; commentary on Āryabhaṭīya, kālakriyāpāda i.5, p. 183.

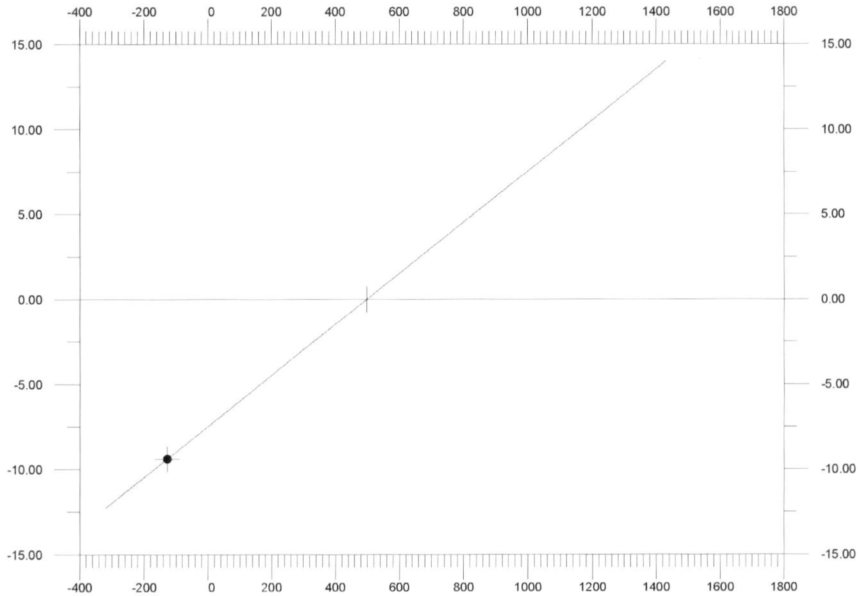

Figure 12: Precession according to the Sūrya Siddhānta, including the Hipparchian origin at -127. The curve passes through zero at the epoch date of Āryabhaṭa's canon, 499 Mar 21.

title. The Latin work is known under the title *De motu octave spere*, and was for a long time falsely attributed to Thābit ibn Qurra. Both of these works present models of accession and recession, equivalent to the Sanskrit *ayanāṃśa*, which is the equation to be applied to the sidereal longitudes of Sun, Moon and planets in order to recover their tropical longitudes.

In both versions the model arises from special assumptions about the longitudes of the stars attributed to Hipparchus, Ptolemy, al-Battānī and al-Zarqālī himself. We recognize that Ptolemy's stellar longitudes are roughly one degree too small for the date adopted by him, but this error was not appreciated by al-Zarqālī, who adopted the interval 2;40 as the increment in longitudes from Hipparchus to Ptolemy. As a result of this error the adopted increments support the postulated model of non-linear increment in longitude, a model which is represented by him as an epicycle.

In the work which we refer to as 'Tract on the Movement of the Fixed Stars' al-Zarqālī presented three versions of this epicycle model, which varied according to the way the epicycle was placed in relation to the equator and ecliptic. In this tract he presented also a separate model designed to agree with the small variations in the obliquity of the ecliptic, supposed to vary periodically around a mean value 23;43. This variation was included in the calculations made from the epicycle models.

A full account will be given elsewhere but here I indicate some of the results. Let the accession be written in general as $E = F(\theta, r, \epsilon)$, where θ is the angle in the epicycle, r is the radius of the epicycle, and ϵ is the obliquity of the ecliptic (which varies with the time). The radius has to be determined, as well as the period P of rotation. In two of the models the obliquity enters into the geometry. It is assumed that this formula will be fitted to four values of the longitude of Regulus (α Leo). Here al-Zarqālī assumes that the years and increments of the four observations are as follows.

	Year	Interval years	Increment in longitude
Hipparchus	-146		
Ptolemy	139	285	2;40
al-Battānī	883	744	11;36
al-Zarqālī	1075	192	2;34

Of course we understand from the *Almagest* that the interval from Hipparchus to Ptolemy is in fact 265 years, not 285. These data have to be fitted by four instances of the relation $E = F(\theta, r, \epsilon)$. There are four unknowns: the values of E and θ at the first year -146, and the radius and period of the epicycle, r and P in years. These conditions are written

$$E + e[n] = F(\theta + I[n] \times 2\pi/P, r, \epsilon[n]), \text{ where } n = 1 \text{ to } 4.$$

Here the dependence on the radius and the obliquity is made explicit. The increments $e[n]$ and $I[n]$ are

$$\begin{array}{llll} I[1] = 0 & I[2] = 285 & I[3] = 1029 & I[4] = 1221 \\ e[1] = 0 & e[2] = 2;40 & e[3] = 14;16 & e[4] = 16;50. \end{array}$$

In contrast to the use of epicycle models in the theories of Sun, Moon and planets, there is here no direct empirical basis for either the period or the radius of the epicycle, so that these have to be determined from this exercise. These general formalities are the same for each of the models, except that in the second the obliquity plays no role.

If this model is to be seen as a continuation of Indian models, then from the knowledge of those models we would expect that the accession is zero some time in the sixth century. In the present model we can determine the value of θ, and hence the date, when $E = 0$, but we cannot do that until the four parameters that fix the model have been discovered by solving the four equations.

Model I

The first of the models results in the formula

$$\sin E = \sin\phi / \sin\epsilon, \text{ where } \tan\phi = \frac{\sin\theta}{A+\cos\theta}, \; A = 1/\sin r$$

whis ϕ an intermediate angle. Therefore in order to fit the model to the assumed conditions we have the four equations $F[n] = 0$, to be solved for the root (E, A, M, θ):

$$F[n] = \sin(E + e[n]) - \sin(\phi[n]) / \sin(\epsilon[n]),$$

$$\tan(\phi[n]) = \frac{\sin(\theta+M\times I[n])}{A+\cos(\theta+M\times I[n])}, \; n = 1, 2, 3, 4.$$

The expressions are here written more clearly in terms of A and M, but from the solution the period will be $P = 2\pi/M$, and the radius will be $r = \arcsin(1/A)$. The solution offered by al-Zarqālī is[36]

$r = 3;54$ $P = 3792$		
	E	θ
Hipparchus	-9	-72;48
Ptolemy	-6;38	-45;45
al-Battānī	4;18	27;15
Zarqala	6;45	46;5

These values are to be sure only an approximation, and if one attempts to determine the year of zero accession it is quickly discovered that they are not consistent. For, from -45;45 in the year 139 we derive 620.9 as the zero year, while from 27;15 in the year 883, we have 596.97.[37]

It is natural for us to determine the root of a set of four equations by an application of the Newton-Raphson iteration.[38] This is an iteration procedure which starts at a point near the root and which leads to an approximation, as

[36]Millás Vallicrosa (1950), p. 316.

[37]The reading of these numbers from the Hebrew text, of which a facsimile is provided, appears to be quite secure.

[38]Press (1992), for the Newton-Raphson method, p. 362; for the inversion of the Jacobian matrix, p. 43. One could equally well use the root-finding algorithms of the GSL library, http://www.gnu.org/software/gsl/.

close as one wishes, to the the exact value. Here is the result of iteration starting with the above approximation.

$r = 4;0,32$ $P = 3505.05$			
	E	ϕ	θ
Hipparchus	-9;24,21	-3;47,24	-74;46,42
Ptolemy	-6;44,21	-2;43,19	-45;30,22
al-Battānī	4;51,39	1;56,26	30;54,34
Zarqala	7;25,39	2;57,45	50;37,46

The zero accession year is now consistently 582.06, apart from multiples of the period P.

Al-Zarqālī gives no serious explanation of his procedure, which he would surely have done if he had been able to proceed systematically. The wonder is that he came as close as he did to fairly good values, which must have involved a serious effort, with much calculation.

However the problem presented by these four equations is solved not merely by fixing this root with improved precision. There is much more to discover, for this is not the only root, far from it. In fact there is an infinity of roots of these four equations. If we carry out even a partial survey of the root space we find a bewildering complexity. There is, for example, the solution

$r = 23;36,14$ $P = 36475.56$		
	E	θ
Hipparchus	-108;23,58	-129;8,18
Ptolemy	-105;43,58	-126;19,32
al-Battānī	-94;7,58	-118;58,57
al-Zarqālī	-91;33,58	-117;5,16

In this case the zero accession year lies far in the future at 12938, or in the past, in the year -23537.

Each of these two roots has a near neighbour. Adjacent to the root at 3505 years is another with the period 4132 years, which was also discovered by al-Zarqālī(who assigns it the period 4692 years) when he looks at the variant of the model in which the angle θ is reckoned from 180°. Adjacent to the root at 36475 is another root at 29548.

When an attempt is made to achieve a broader survey of the root space it is soon found that apart the pairs of roots just mentioned, which occur symmetrically in each of the four quadrants, there is in the close neighbourhood of

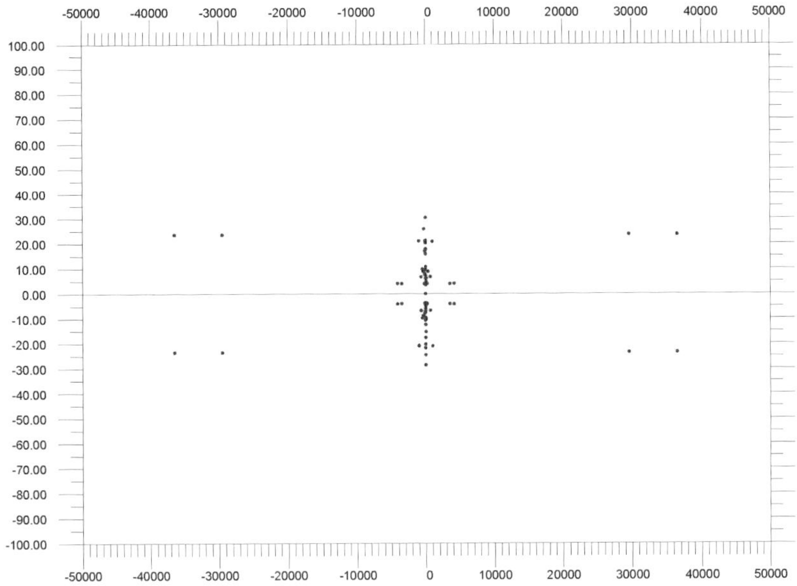

Figure 13: Roots of the system $F[n] = 0$, plotting radius *vs* period, in one two dimensional plane of the four dimensional root space. Each of the roots is found by Newton-Raphson iteration

the root axis an infinity of roots. Some idea of this can be seen in Figure 13, which shows the roots in one particular cross section of the four dimensional root space. In this plot one can see the pair of roots near the period 4000, already known to al-Zarqālī, and another pair near the period 30000.

It would take us far beyond the present scope to discuss the four dimensional region surrounding each root, corresponding to the 'basin of convergence' familiar from studies of fractals.

As remarked above three models were considered by al-Zarqālī, of which he decided to prefer the third. From the present point of view there is in fact little to choose, since all have a great multiplicity of roots, and always including one that is a match for the Indian scheme of accession.

It is impossible to say whether al-Zarqālī had any notion of the multiplicity of roots, though it would seem unlikely. He has found in any case a root which fits well with such information as he had from Indian sources. Of these he names the *Sindhind*, which is the version of the *Brāhmasphuṭasiddhānta* known to the Arabic world, although the Sanskrit original, at least, gives no version of accession and recession. He certainly assumed that he should obtain a version of accession that vanishes around the 582, for as he says[39]

> Certainly when the head of Aries is found on the equatorial circle, the head of Aries and the vernal point coincide exactly, as happened forty years before the Hijra, that is at the birth of the Prophet of God [A.D. 622]

De motu octave spere

In the version of accession and recession which is preserved in this Latin text, and which has been studied rather extensively, there is a model which leads to a joint variation of both obliquity and precession. Here, as I have argued elsewhere, al-Zarqālī began with tropical Sun of the zīj of al-Battānī.[40] Taking the synodic differences of that zīj, he then added a sidereal Sun which led to a new canon with Sidereal longitudes, a canon which was then known as the Toledan Tables. He then created a model of precession which, when applied to the sidereal Sun of this canon, would yield a tropical Sun that is reconciled with the observations of the Sun by Hipparchus, Ptolemy, and those made by himself around 1075. This version of precession vanishes in the year 604. This approach to the problem of precession is then totally different from that in the Tract on the Motion of the Fixed Stars, but it again depends crucially on the Ptolemaic errors in the Sun and in the fixed stars, errors which of course come to the same thing in the end.

Sidereal ecliptics: Indian and Arabic

When precession is incorporated in an Indian canon it is always (to my knowledge) assumed to be periodic, a cycle of accession and recession. This owes something, at least, to the procedures of Greek astrology, such as Theon's well known model, but not, surely, to any attempt to fit a model to the star coordinates adopted by Ptolemy, such as al-Zarqālī does here.

The sidereal coordinates depend on a conventional origin point in the ecliptic, a point marked by a star, from which the longitudes were measured. This is a convention and any star will do for this purpose. The origin point assumed by the Indians was marked by the star ζ Psc, as noted above, and

[39] Millás Vallicrosa (1950), p. 338.
[40] Mercier (1996).

appears to have been fixed by Āryabhaṭa, who decided in effect to initiate a new convention of sidereal longitude.

Al-Zarqālī determines a model of precession, and his effort depends on a choice of the convention of sidereal longitude. Where, in his work, does he define that longitude? In what way does he indicate the origin point among the stars? One searches in vain for any explicit indication of how he defines this, and the nearest he comes to it is in a remark where, in defining the arrangement of longitude determinations at the various times, he says, 'We suppose there is the point D whose position in the sphere of the fixed stars is known to us, and which represents the head of Aries.'[41] This position is left however to be determined by the solution of his epicycle model, and this in turn depends on precession data that included the erroneous star coordinates of Ptolemy. While the Indian choice of sidereal longitude was dictated by Āryabhaṭa's foundation of his canon, and had nothing to do with the erroneous character of Ptolemy's star longitudes, al-Zarqālī's determination depends essentially on that erroneous character.

The solution of the problem as al-Zarqālī has defined it ends with the determination of the origin of the sidereal longitudes. That is to say his determination of the origin of sidereal longitudes is a consequence of his use of an epicycle model together with his adoption of Ptolemy's star coordinates. The longitude of Regulus according to Ptolemy is 122;30, but according to the above solution the accession at his time is -6;44,21 (by the corrected result), so that when the accession is zero in the year 582 the sidereal and tropical longitudes coincide, then the longitude of Regulus will be 129;14,21. That means that the origin of sidereal longitudes is at the point East of Regulus by the amount, 129;14,21, at least according to the Ptolemaic coordinate of the star. In that year in fact the longitude of the star ζ Psc is close to -0;15, so that we have arrived at the origin of the Indian sidereal longitude. Such an origin point is altogether a matter of convention, and if this was the convention that suited the Indian astronomers, why should it be right for al-Zarqālī? As just remarked, he never provides any definition himself of the convention of sidereal longitude.

The irony of course is that if instead of Ptolemy's position of Regulus, 122;30, he had adopted a more correct value such as 124, for the year 139, which would entail a more or less linear precessional increase from Hipparchus to his own time, then he could never have fitted any sort of epicycle,

[41]Millás Vallicrosa (1950), p. 301, 'Supongamos el punto D, cuya posición en las esfera de las estrellas fijas nos es conocida, el cual representa el principio de Aries.'

and the whole program would have collapsed, without any determination of sidereal longitudes. His result, in which he arrives at a sidereal origin, thus depends not only on bad data from Ptolemy's star list, but on the choice of one of a great multiplicity of roots of the epicycle model which is used to fit this bad data. The result is thus unreal from the point of view of both astronomical data and astronomical convention: such is the futility of the whole exercise.

References

Bhandarkar (1983). 'A List of the Inscriptions of Northern India in Brahmi and its derivative Scripts, from about 200 A.C.', D.R. Bhandarkar, Appendix to *Epigraphia Indica*, Volumes XIX to XXIII, Archaeological Survey of India, 1983.

Billard (1956). 'Perspectives nouvelles sur l'astronomie indienne', Roger Billard, *Artibus Asiae*, XIX, 3/4 (1956), pp.186-196.

Billard (1971). *L'Astronomie indienne. Investigation des textes sanskrits et des données numériques.* Publications de l'École française d'Extrême-Orient, Volume LXXXIII, Paris, 1971.

Fleet (1876). 'Sanskrit and Canarese Inscriptions', J.F. Fleet, *Indian Antiquary* V(1876) p. 67-76.

Fleet (1911). 'The Kaliyuga Era of B.C. 3102', J. F. Fleet, *Journal of the Royal Asiatic Society* 1911, p. 675-698.

González-Reimann (2002). *The Mahābhārata and the Yugas*, Luis González-Riemann, New York: Peter Lang, 2002.

Kern (1874). *The Âryabhatîya, a manual of astronomy, with the Commentary Bhatadîpikâ of Paramâdîçvara*, edited by H. Kern, Leiden: Brill 1874.

Kielhorn (1891). 'Examination of questions connected with the Vikrama Era', *Indian Antiquary*, January 1890 - December 1891.

Kielhorn (1897). 'On the Dates of the Saka Era in Inscriptions' *Indian Antiquary*, May 1894 -June 1897.

Kielhorn (1899). 'A List of the Inscriptions of Northern India from about A.D. 400', F. Kielhorn, Appendix to *Epigraphia Indica* Vol V, 1899.

Kielhorn (1903). 'A List of the Inscriptions of Southern India from about A.D. 500', F. Kielhorn, Appendix to *Epigraphia Indica* Vol VII, 1903.

Kramrisch (1928). The *Vishnudharmottara (Part III)*, Stella Kramrisch, Second Revised and Enlarged Edition, Calcutta: University Press, 1928.

Matsya Puranam, translated by 'A Taluqdar of Oudh', The Sacred Books of the Hindus, Vol. XVIII, in two parts, Allahabad, 1916.

Mercier (1976-7). 'Studies in the Medieval Conception of Precession', Raymond Mercier, *AIHS* 26 (1976) 197-220, 27 (1977) 33-71. Reproduced in Mercier (2004).

Mercier (1987). 'The meridians of reference of Indian astronomical canons', Raymond Mercier, in G. Swarup, A.K. Bag & K.S. Shukla eds., *History of Oriental Astronomy, Proc of IAU Colloquium 91*, New Delhi, 13-16 Nov 1985. Cambridge: University Press: 97-107. Reproduced in Mercier (2004).

Mercier (1996). 'Accession and recession: reconstruction of the parameters', Raymond Mercier, in *From Baghdad to Barcelona*, edited J. Casulleras and J. Samsó, two volumes, Barcelona 1996; vol. 1, pp. 299-347. Reproduced in Mercier (2004).

Mercier (2004). *Studies on the Transmission of Medieval Mathematical Astronomy*, Raymond Mercier, Variorum Collected Studies Series CS787, Ashgate Variorum, Ashgate: Aldershot Hants, 2004. ISBN 0-86078-949-7.

Mercier (2011). 'On the originality of Indian Mathematical Astronomy', Ch. 8.e, in *Interactions of Ancient Astral Science*, Raymond Mercier, Vergleichende Studien zu Antike und Orient, X, edited by David Brown, Bremen: Hempen, 2011.

Millás Vallicrosa (1950). *Estudios sobre Azarquiel*, José Millás Vallicrosa, Madrid-Granada: Consejo Supérior de Investigaciones Científicas, 1943-1950.

Pargiter (1918). *The Purana Text of the Dynasties of the Kali Age*, with Introduction and Notes by R.E. Pargiter, Oxford: University Press.

Pillai (1957). *Āryabhatīyam*, with commentary by Nīlakaṇtha Somayaji, ed. S. K. Pillai, University of Travancore Sanskrit Series Nos 101, 110, 185, Trivandrum 1930-1957.

Pingree (1968). 'The Pitāmahasiddhānta of the Viṣṇudharottarapurāṇa', translated by David Pingree, The Adyar Library Bulletin Vols. XXXI-XXXII, 1967-8.

Pingree (1976). 'The recovery of early Greek astronomy from India.' David Pingree, *Journal for History of Astronomy* 7: 109-23.

Pingree (1981). 'History of Mathematical Astronomy in India', David Pingree, *Dictionary of Scientific Biography*, Volume XV Supplement I, New York: Charles Scribner's, 1981.

Plofker (2009). *Mathematics in India*, Kim Plofker, Princeton, 2009.

Press (1992). *Numerical Recipes in C*, William H. Press, Saul A. Teukolsky, William T. Vetterling, Brian P. Flannery, Cambridge: University Press, 1992.

Renou (1947). *L'Inde classique, Manuel des études indiennes*, Louis Renou, Jean Filliozat, 2 vols., Paris : Payot.

Sastry (1993). *Pañcasiddhāntikā of Varāhamihira* with translation and notes by T.S. Kuppanna Sastry, critically edited with introduction and appendices by K.V. Sarma, P.P.S.T. Foundation [Patriotic and People Oriented Science and Technology Foundation], Adyar, Madras, 1993. P.P.S.T. Science Series No. 1.

Shah (2002). *Visnudharmottara-Purana*: Translated into English from Original Sanskrit Text, Parimal Pub, 1999 (vol.I), 2002 (vol.II & III).

Shukla, K. S. 1976. *Āryabhaṭīya of Āryabhaṭa with the commentary of Bhāskara I and Someśvara*, critically edited by Kripa Shankar Shukla, Three Parts, Delhi: Indian National Science Academy, 1976.

Sircar (1983). *Select Inscriptions bearing on Indian History and Civilisation from the sixth to the Eighteenth Century* A.D., Edited by Dines Chandra Sircar, Delhi: Motilal Banarsidas, 1983.

van Buitenen (1975). *The Mahābhārata. 2 The book of the assembly Hall, 3 The Book of the Forest*, translated and edited by J.A.B. van Buitenen, Chicago: The University Press, 1975.

Verbrugghe (2001). *Berossos and Manetho, Native Traditions in Ancient Mesopotamia and Egypt*, Gerald P. Verbrugghe and John M. Wickersham, Ann Arbor: University of Michigan Press, 2001.

Entre astronomie et mathématiques, les découvertes indiennes en trigonométrie : la construction des tables de sinus et quelques méthodes d'interpolation

Jean Michel Delire
Centre ALTAÏR d'histoire des sciences et des techniques de l'ULB

Avant de rentrer dans le vif du sujet, rappelons que la notion de *sinus* est d'origine indienne. Elle s'oppose aux cordes des Grecs et porte le nom d'ardhajyā ou jyārdha 'demi-corde', ou, plus simplement jyā. Le terme jyā, qui prend parfois la forme jīva/jība, transmis aux savants arabo-musulmans, a été retranscrit j-y-b (جيب), prononcé jīb, mais aussi jaib[1], 'poche' en arabe, avant d'être traduit par le terme latin correspondant '*sinus*'. Notons que (ardha)jyā(rdha) désigne le *sinus* dans un cercle de rayon r, donc plutôt le $rsinus$[2] que notre *sinus* moderne.

La première table de *rsinus*

Les premiers traités astronomico-mathématiques indiens, dits siddhāntas, sont perdus sous leur forme originale, mais ils ont été résumés par Varāhamihira dans son *Pañcasiddhāntika* (VIe siècle). Le plus ancien d'entre eux contenant une table de $rsinus$ est le *Pauliśasiddhānta*, peut-être d'origine grecque comme son nom l'indique. Tabulés de 3°45' en 3°45' dans le premier quadrant, ces $rsinus$ sont exprimés sous forme sexagésimale (voir table 1) pour un cercle de rayon $r = 120$.

[1]Du fait de l'absence de notation des voyelles, d'après P.-S.Filliozat, « *Calculs de demi-cordes par Āryabhaṭa et Bhāskara I* », *Bulletin d'études indiennes*, 6 (1988), pp.255-274.

[2]Bien entendu, on a $rsin\alpha = r.sin\alpha$ ou $r.sin\ (\alpha/r)$, si l'on mesure l'arc en radians.

Table 1

N°	Angles (en ° ')	rsinus (*P.S.*#)	Différences (*Āryabhaṭīya*)	rsinus (*S.S.*)	Différences du 2nd ordre	rsinus verses
1	3°45'	7 ;51	225'	225'		7'
2	7°30'	15 ;40	224'	449'	- 1'	29'
3	11°15'	23 ;25	222'	671'	- 2'	66'
4	15°	31 ;04	219'	890'	- 3'	117'
5	18°45'	38 ;34	215'	1105'	- 4'	182'
6	22°30'	45 ;56	210'	1315'	- 5'	261'
7	26°15'	53 ;05	205'	1520'	- 5'	354'
8	30°	60	199'	1719'	- 6'	460'
9	33°45'	66 ;40	191'	1910'	- 8'	579'
10	37°30'	73 ;03	183'	2093'	- 8'	710'
11	41°15'	79 ;07	174'	2267'	- 9'	853'
12	45°	84 ;51	164'	2431'	- 10'	1007'
13	48°45'	90 ;13	154'	2585'	- 10'	1171'
14	52°30'	95 ;12	143'	2728'	- 11'	1345'
15	56°15'	99 ;46	131'	2859'	- 12'	1528'
16	60°	103 ;55	119'	2978'	- 12'	1719'
17	63°45'	107 ;37	106'	3084'	- 13'	1918'
18	67°30'	110 ;52	93'	3177'	- 13'	2123'
19	71°15'	113 ;37	79'	3256'	- 14'	2333'
20	75°	115 ;55	65'	3321'	- 14'	2548'
21	78°45'	117 ;42	51'	3372'	- 14'	2767'
22	82°30'	118 ;59	37'	3409'	- 14'	2989'
23	87°15'	119 ;44	22'	3431'	- 15'	3213'
24	90°	120	7'	3438'	- 15'	3438'

D'après M.Yano, « *Three Types of Hindu Sine Tables* », *IJHS*, 12 (1977), 83-89, qui se base sur l'édition de Neugebauer-Pingree, 1970-1. Mais G.Mazars, « *La notion de sinus dans les mathématiques indiennes* », *Fundamenta Scientiae*, 15 (1974), pp.1-23, affirme que le chapitre IV du *Pañcasiddhāntika* ne serait pas attribué. Et pourtant, G.Mazars utilise la même édition que M.Yano et cite les méthodes de ce chapitre sous les mêmes numéros (IV.1-5).

Partant des $rsinus$ de 90° (n°24) et 30° (n°8), respectivement égaux à r et $r/2$, le *Pauliśasiddhānta* calcule les autres valeurs par complémentation et dimidiation :
Dimidiation :

$$rsin\frac{\alpha}{2} = \sqrt{\left[\frac{r - rsin(90° - \alpha)}{2}\right]^2 + \left[\frac{rsin\alpha}{2}\right]^2}$$

équivalent moderne de *PS* IV, 3-4, simplifiée ensuite en

$$rsin\frac{\alpha}{2} = \sqrt{60\,[r - rsin(90° - \alpha)]} \quad (\textit{PS}\,\mathrm{IV}, 5)$$

Complémentation :

$$rsin(90° - \alpha) = \sqrt{r^2 - rsin^2\alpha} \quad (\textit{PS}\,\mathrm{IV}, 4)$$

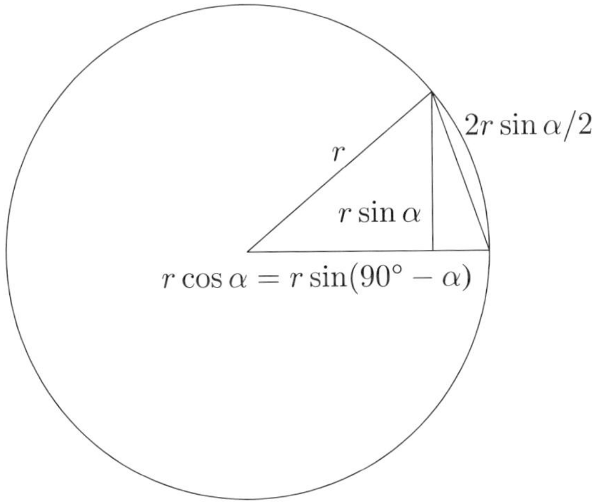

Figure 1

La première identité découle du théorème de Pythagore (figure 1) appliqué au triangle rectangle d'hypoténuse $2rsin\alpha/2$.

La deuxième identité est une autre écriture de la première. Elle est moins générale, mais a l'avantage de simplifier les calculs pour le cas particulier où $r = 120$, puisqu'une multiplication par 60 est aisée en numération sexagésimale.

On utilise alors les séquences suivantes pour calculer les 24 $rsinus$:

Table 2

Dimidiation	→	90°	→	#45°	→	22°30’	→	11°15’	30°	→	15°	→	7°30’	→	3°45’
Complémentation	↓					↓		↓	↓		↓		↓		↓
						↓		78°45’	60°		↓		↓		86°15’
						67°30’		33°45’			↓		82°30’	→	41°15’
								↓			↓				↓
								56°15’			↓				48°45’
											75°	→	37°30’	→	18°45’
													↓		↓
													↓		71°15’
													↓		
													52°30’	→	26°15’
															↓
															63°45’

Il faut noter que $rsin45° = 84;51 = 60.(1;24,51)$ (approximation babylonienne classique de $\sqrt{2}$), ce qui peut aussi être obtenu par dimidiation : $rsin45° = rsin90°/2 = \sqrt{60.120} = \sqrt{7200}$. Par la méthode classique d'extraction (*Āryabhaṭīya*, II.4), $84^2 = 7056$ est le plus grand carré inférieur à 7200 et la division du reste 144 par $2.84 = 168$ donne $51,4\ldots$, soit 51 en arrondi.

Utilité astronomique

Ce genre de table était utile pour certains calculs astronomiques, que nous allons maintenant expliquer. Le modèle de base de l'astronomie siddhantique pour les mouvements planétaires peut être décrit comme un modèle excentrico-épicyclique. La planète se meut sur un cercle (épicycle) dont le centre C' (figure 2) tourne lui-même sur un cercle dont le centre O' est décalé (excentrique) par rapport à la terre O.

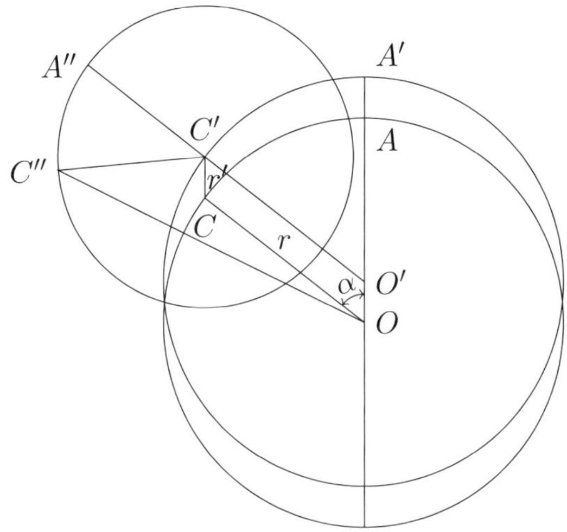

Figure 2

En supposant que C est le quatrième sommet du parallélogramme $OO'C'C$ et que C tourne uniformément autour de O (mouvement moyen), les « corrections » suivantes sont apportées à ce mouvement moyen pour permettre le calcul de la position de la planète :

La correction *manda*[3] : c'est l'angle $C'OC = \mu$, avec[4]

$$sin\mu = \frac{r'sin\alpha}{\sqrt{(r'sin\alpha)^2 + (r + r'cos\alpha)^2}}$$

(α est l'angle AOC, où A est la position initiale, $r = OC = O'C'$ et $r' = OO' = CC'$). Cette correction est due à l'excentricité, mais il est

[3]Sens principal 'lent'.

[4]Note de l'éditeur : voir, ci-dessous, dans l'article de K.Ramasubramanian, p.128, une justification de cette formule.

clair que r' peut être interprété comme le rayon d'un épicycle sur lequel se meut C' (Figure 3). Dans cette interprétation, le modèle devient doublement épicyclique.

la correction *śīghra*[5] : c'est l'angle $C'OC'' = \sigma$, avec

$$sin\sigma = \frac{r''sin\varphi}{\sqrt{(r''sin\varphi)^2 + (R + r''cos\varphi)^2}}$$

(où $\varphi = \mu + \gamma$, $\gamma = \lambda_S - \alpha$ et $R = \sqrt{(r'sin\alpha)^2 + (r + r'cos\alpha)^2} = C'O$)

Dans la situation représentée par la Figure 3 : α (qui caractérise le mouvement moyen de C) est donc corrigé en $\alpha'' = \alpha - \mu + \sigma$ en vue de calculer l'angle α'' dans la direction duquel doit se trouver la planète, par rapport à sa position initiale. Bien sûr, tous ces angles dépendent finalement de α et λ_S. Ceux-ci sont calculés, à partir d'une conjonction initiale des planètes et du soleil qui se serait produite le 18 février -3101, à l'aide des périodes solaire et planétaires moyennes[6] et du nombre de jours, l'*ahargaṇa*, écoulés depuis la conjonction. Nous ne donnerons pas ici les détails de ces calculs. Ce qu'il est important de souligner, c'est la nécessité des fonctions et tables trigonométriques dans les traités astronomico-mathématiques indiens.

La table de différences de rsinus d'Āryabhaṭa

Āryabhaṭa, lui aussi, donne une table de $rsinus$ dans le premier chapitre (astronomique) de son *Āryabhaṭīya*.[7] Il s'agit plutôt d'une table de différences de $rsinus$, établie pour un cercle de rayon $r = 3438$ (table 1, colonne 4). La raison de ce choix pour r est la suivante : on cherche à obtenir un cercle dont le périmètre égale le nombre de minutes correspondant à un tour complet, soit $360 \times 60 = 21600$. Le rayon de ce cercle est donc

[5]Sens principal 'rapide'.

[6]Données, par exemple, en *Āryabhaṭīya*, I.3-4. Voir *Āryabhaṭīya of Āryabhaṭa*, Critically edited with Introduction, English Translation, Notes, Comments and Indexes, by K.S.Shukla, In collaboration with K.V.Sarma, Indian National Science Academy, New Delhi, 1976, pp.6-8. (Note de l'éditeur : voir déjà l'article de R.Mercier, ci-dessus, p.16, et celui de K.Ramasubramanian, p.134.)

[7]*Āryabhaṭīya of Āryabhaṭa*,..., p.29. Signalons qu'*Āryabhaṭa* nomme kalārdhajyā 'demi-cordes des seizièmes [de quadrant]' cette table, terme dont pourrait provenir le terme 'arabe' kar(a)dajāt désignant les tables de $sinus$ inspirées des tables indiennes (v. aussi F.Woepcke, « *Sur le mot kardaga et sur une méthode indienne pour calculer les* $sinus$ », *Nouvelles Annales de Mathématiques*, 13 (1854), 386-394).

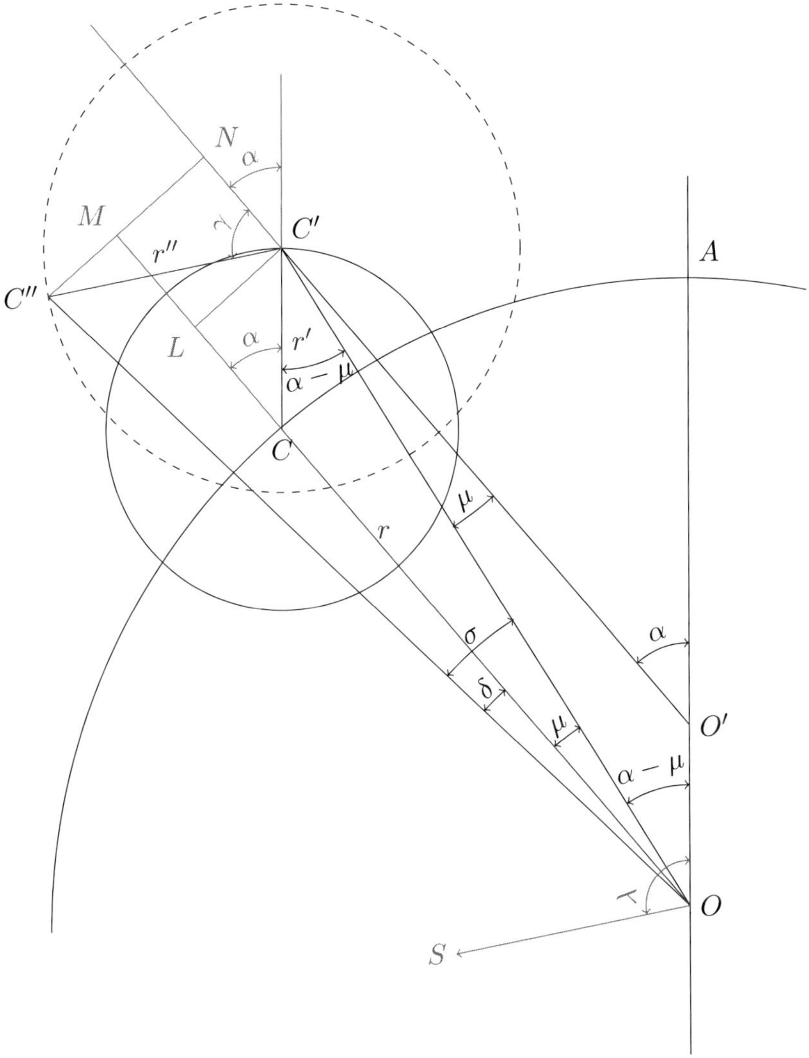

Figure 3

$21600/2\pi \approx 21600/6,2832 \approx 3438$.[8] L'un des siddhānta décrits par le *Pañcasiddhāntika*, le *Sūryasiddhānta*,[9] présente la table de $rsinus$ (table 1, colonne 5) correspondant à la table de différences de l'*Āryabhaṭīya*.

Dans son deuxième chapitre (nommé *gaṇita* 'mathématique'), Āryabhaṭa décrit une méthode permettant d'obtenir ces $rsinus$:

samavṛttaparidhipādam chindyāt tribhujāccaturbhujāccaiva /
samacāpajyārdhāni tu viṣkambhārdhe yatheṣṭāni //11//

« *si l'on divise un quart de circonférence de cercle [en arcs égaux, on peut trouver] autant de* $rsinus$ *d'arcs égaux qu'on désire sur le demi-diamètre à partir de triangles et de quadrilatères.* »[10]

Il pourrait s'agir d'une méthode similaire à celle utilisée par le *Pauliśa-siddhānta*, et c'est bien ainsi que l'entend Bhāskara I dans son commentaire (629) de l'*Āryabhaṭīya*.[11]

Mais notre traduction suggère que II.11 indique simplement un découpage[12], à partir duquel la règle suivante de l'*Āryabhaṭīya* est d'application[13] :

prathamāccāpajyārdhādyairūnam khaṇḍitam dvitīyārdham /
tatprathamajyārdhāṁśaistaistairūnāni śeṣāṇi //12//[14]

« *à partir du premier, le deuxième* $rsinus$ *[est] diminué [du] quotient [par le premier] d[u](es)* $rsinus$ *de(s) [l']arc(s) précédent(s),*
les autres [$rsinus$ *sont] diminués de chacun des quotients des* $rsinus$ *[précédents divisés] par le premier.* »

---

[8]*Āryabhaṭa* donne à π la valeur approchée $3,1416$ (*Āryabhaṭīya of Āryabhaṭa*, ..., II.10, p.45).

[9]*The Sūrya Siddhānta, A Text-book of Hindu Astronomy*, Translated with Notes and Appendix by E.Burgess (1860), Edited by P.Gangooly, With an Introduction by P.Sengupta, Motilal Banarsidass ed., Delhi, 2000 (1e éd. 1935), p.60. Ebenezer Burgess (1805-1870) était un missionnaire américain, qui vécut en Inde de 1839 à 1854.

[10]*Āryabhaṭīya of Āryabhaṭa*, ..., II.11, p.45. Voir P.-S.Filliozat, *op.cit.*, p.257, pour une traduction différente.

[11]*Āryabhaṭīya of Āryabhaṭa with the Commentary of Bhāskara I and Someśvara*, Critically edited with Introduction and Appendices, by K.S.Shukla, Indian National Science Academy, New Delhi, 1976, pp.77-83 (texte sanskrit) ; A.Keller, *Expounding the Mathematical Seed*, Birkhaüser Verlag, 2006, vol.1, pp.59-67 (traduction anglaise) ; P.-S.Filliozat, *op.cit.*, p.260, sq., pour une explicitation du commentaire de Bhāskara I, selon laquelle il paraît clair que Bhāskara interprète II.11 comme équivalente à celle du *Pauliśasiddhānta*.

[12]Selon P.-S.Filliozat, *op.cit.*, p.258, le commentateur Sūryadeva Yajvan (1191) décrit un découpage en 24 demi-cordes.

[13]Bhāskara I, qui considère II.11 comme suffisante par elle-même, critique II.12 parce qu'elle s'appuierait sur II.11 pour ne rien apporter de plus. Voir P.-S.Filliozat, *op.cit.*, p.269, pour le détail de cette critique.

[14]*Āryabhaṭīya of Āryabhaṭa*, ..., II.12, p.51.

Ce qui signifie, en symboles modernes :

$$\begin{aligned}\delta_1 &= rsinh\\ \delta_2 &= rsinh - \frac{rsinh}{rsinh} = rsinh - 1\\ \delta_{n+1} &= rsinh - \frac{rsinh + rsin2h + \ldots + rsin\,nh}{rsinh} = \delta_n - \frac{rsin\,nh}{rsinh},\end{aligned}$$

où $h = 225' \approx r\sin h$[15]

Il s'agit donc d'une méthode itérative, permettant d'obtenir chacune des différences de $rsinus$ à partir des précédentes.

Le *Sūryasiddhānta* donne une règle équivalente[16] et les historiens modernes, son éditeur Burgess le premier[17], suggèrent qu'elle a pu être obtenue en calculant les différences secondes (voir table 1, colonne 6) et en observant qu'elles sont proportionnelles aux $rsinus$ eux-mêmes. Ainsi :

$$\begin{aligned}\Delta 1 &= \delta_2 - \delta_1 = -1 = k \cdot rsinh\\ \Delta 2 &= \delta_3 - \delta_2 = -2 = k \cdot rsin2h\\ \Delta n &= \delta_{n+1} - \delta_n = k \cdot rsin\,nh, \text{avec } k = -\frac{1}{rsinh} \approx -\frac{1}{225}\end{aligned}$$

Cette valeur de k est inexacte, bien entendu, mais son choix découle de la valeur donnée à r (3438) et du désir de donner aux différences de $rsinus$, et aux différents $rsinus$, des valeurs entières. En fait, Delambre[18] a calculé une valeur plus correcte $k \approx -\frac{1}{233.5}$ à l'aide de l'identité $\Delta n = \delta_{n+1} - \delta_n = 2(cosh - 1).rsin\,nh$,[19] pour $h = 3°45'$.[20]

[15]En ce qui concerne le choix de $3°45'$ pour h, petit angle de base pour lequel $rsinh \approx h$, et la croyance de certains mathématiciens-astronomes indiens en l'existence d'une corde égale à un arc, il faut encore citer le commentaire de Bhāskara I : « Nous, nous disons qu'il existe une corde égale à un arc. S'il n'y en avait pas, alors sur la terre égale il n'y aurait pas stabilisation d'une boule de fer. Nous inférons donc qu'il existe un tel espace du cercle, par quoi sur la terre égale la dite boule de fer se stabilise. Et cet espace est la quatre-vingt seizième partie de la circonférence » (traduction de P.-S.Filliozat, *op.cit.*, p.260).

[16]II.15-16, in *The Sūrya Siddhānta*, ..., p.58.

[17]*The Sūrya Siddhānta*, ..., p.61.

[18]Jean-Baptiste Delambre, *Histoire de l'astronomie ancienne*, Paris, 1817, Vol.1, p. 458.

[19]Il s'agit d'une 'simple' application des formules de Simpson :
$\delta_{n+1} - \delta_n = [rsin(n+1)h - rsin\,nh] - [rsin\,nh - rsin(n-1)h] = rsin(n+1)h + rsin(n-1)h - 2rsin\,nh = 2rsin\,nh.cosh - 2rsin\,nh = 2rsin\,nh.(cosh - 1)$.

[20]Avec $cos3°45' \approx 0,9978589$, on a $2(cosh-1) \approx 2(0,9978589-1) \approx -0,0042822 \approx$

La notion de preuve dans les mathématiques indiennes

Jusqu'ici, la méthode itérative se présente comme une recette découlant, ainsi que le suggérait Burgess, d'une « simple » observation de la proportionnalité existant entre les *sinus* tabulés et leurs différences secondes. La recette fonctionne, donc les mathématiciens astronomes indiens sont satisfaits et aucun d'entre eux ne se préoccupe d'investiguer le pourquoi de ce bon fonctionnement. C'est souvent sous cette forme caricaturale qu'ont été présentées les méthodes mathématiques de l'Inde ancienne. Pourtant, il existe une forme d'investigation, si pas de preuve, de la plupart des méthodes utilisées. C'est la notion d'upapatti. L'upapatti n'est pas une preuve, au sens hypothético-déductif du terme. Les mathématiciens indiens ne cherchent pas, au contraire de leurs collègues grecs ou descendants de ceux-ci, à construire un système de vérités mathématiques à partir d'un ensemble d'axiomes. Il s'agit plutôt « (...) de convaincre l'étudiant intelligent de la validité d'un théorème de sorte qu'une démonstration visuelle était une forme largement acceptée de preuve (...) »[21]. Selon Bhāskara II, « Sans la connaissance des upapatti, (...) un mathématicien n'acquerra pas son statut dans les assemblées savantes ; sans les upapatti, il ne sera pas lui-même libéré du doute. »[22]. Le commentateur de Bhāskara II, Gaṇeśa Daivajña (né en 1507), précise : « Quoi qu'on établisse dans les sections vyakta ou avyakta[23] des mathématiques, sans upapatti, ce ne sera pas libre de confusion (nirbhrānta) ; cela n'acquerra pas de statut dans une assemblée de mathématiciens. L'upapatti peut être perçu directement comme un miroir dans la main. C'est pourquoi, ainsi que pour l'élévation de la compréhension (buddhivṛddhi), je m'apprête à énoncer les upapatti dans leur intégralité »[24]. On voit que la notion d'upapatti a pour but d'ôter le doute, de convaincre, mais aussi qu'elle est tributaire

---

$-1/233,5248237$. Notons que Nilākaṇṭha (c.1500, voir ci-dessous) a proposé $233,5'$, la même valeur que Delambre, alors que Śaṅkara (milieu du 16e siècle) l'a améliorée en $233'32''$.

[21]T.A.Saraswati Amma, *Geometry in Ancient and Medieval India*, Varanasi, 1979, p.3.

[22]*Siddhāntaśiromaṇi*, Muralidhara Chaturveda (éd.), Varanasi, 1981, p.326, d'après M.D.Srinivas, in *Contributions to the History of Indian Mathematics* (éd.G.G.Emsch, R.Sridharan, M.D.Srinivas), New Delhi, 2005, p.228.

[23]vyakta 'manifesté, apparent' et avyakta 'non manifesté, inconnu (en math.)'. Il s'agit donc des sections algébrique (avyakta ou aussi bījāgaṇita) et non algébrique (vyakta ou aussi pāṭīgaṇita 'calcul sur la planche') de l'arithmétique (mais aussi d'autres domaines, où cette distinction a un sens).

[24]Commentaire *Buddhivilāsinī* (1545) à la *Līlāvatī*, V.G.Apte (éd.), vol.I, Pune, 1937, p.3, d'après M.D.Srinivas, *op.cit.*, p.229.

de l'audience, comme la preuve de type hypothético-déductive elle-même. On approche là un point sensible, aujourd'hui encore, de l'enseignement des mathématiques : « prouver ou convaincre ? »[25]. Il faut encore insister sur le fait que, si les histoires occidentales des mathématiques actuelles[26] ont tendance à passer sous silence cette forme indienne de « preuve », ce n'était pas le cas des premiers chercheurs dans ce domaine, comme l'illustre Henry Thomas Colebrooke[27] (1765-1837, en Inde de 1783 à 1815) ou Charles M.Wish[28].

Les mathématiciens astronomes du Kerala, successeurs d'Āryabhaṭa

Nous allons maintenant voir, à propos des tables de $rsinus$ établies à l'aide d'une méthode itérative, un exemple d'upapatti développé dans les œuvres des mathématiciens du Kerala.

Se présentant généralement comme les successeurs d'Āryabhaṭa[29], dont ils ont souvent commenté les œuvres, les mathématiciens-astronomes kéralais principaux sont :

Nilākaṇṭha Somayāji (1444-c.1545), un brahmane Nampūtiri qui vécut dans le village de Tṛkkaṇṭiyur, proche de Tirur, un lieu de grande réputation dans le domaine du savoir. Son *Tantrasaṅgraha*[30], composé en 1500, est

[25]Formulation proche du titre d'un article de Nicolas Rouche, « *Prouver : amener à l'évidence ou contrôler des implications ?* », in *La démonstration mathématique dans l'histoire* (éd. E.Barbin), IREM de Besançon, 1989.

[26]A l'exception notable de G.G.Joseph, qui aborde le sujet dans son *The Crest of the Peacock*, Penguin Books, 2000 (2e édition).

[27]*Algebra with Arithmetic and Mensuration from the Sanscrit of Brahmegupta and Bhascara*, Londres, 1817.

[28]« *On the Hindú Quadrature of the Circle, and the Infinite Series of the proportion of the circumference to the diameter exhibited in the four S'ástras, the Tantra Sangraham, Yucti Bháshá, Carana Padhati, and Sadratnamála* », By Charles M.Whish, Esq., of the Hon. East-India Company's Civil Service on the Madras Establishment, *Transactions of the Royal Asiatic Society*, vol.3 (1835), pp.509-523. Note de l'éd. : on trouvera dans l'article de K.Mahesh et al. (p.175) plus de détails sur les méthodes kéralaises d'évaluation de π par des séries.

[29]Au point qu'on a parfois prétendu qu'il était originaire du sud de l'Inde (ainsi, K.Sāmbaśivaśāstri, éditeur de l'*Āryabhaṭīya with the Bhāṣya of Nilākaṇṭha Somasutvan*, part.I, Trivandrum, 1930, p.viii), ce que K.V.Sarma (*A History of the Kerala School of Hindu Astronomy*, Vishveshvaranand Institute, Hoshiarpur, 1972, p.6) considère comme non prouvé.

[30]Il existe deux éditions du *Tantrasaṅgraha* : *Tantrasaṅgraha of Nilākaṇṭha Somayāji with Yuktidīpikā and Laghuvivṛti of Śaṅkara*, critically edited by K.V.Sarma, V.V.B.I.S. & I.S., Hoshiarpur, 1977, et *The Tantrasaṅgraha, A Work on Gaṇita, by GārgyaKerala Nilākaṇṭha Somasutvan with Laghuvivṛti of Śaṅkara Vāriar,* published by S.K.Pillai, Trivandrum Sans-

un traité complet et très érudit d'astronomie, en 432 vers, divisés en 8 chapitres. Dans le deuxième chapitre, consacré aux positions vraies des planètes, Nilākaṇṭha traite du problème de la construction d'une table précise de ce que nous appelons aujourd'hui fonctions trigonométriques, *sinus*, *cosinus*, *sinusverse*, etc.[31]

En tant que premier successeur du *Tantrasaṅgraha*, nous prendrons aussi en considération la *Yuktibhāṣā*[32], dont l'auteur Jyeṣṭhadeva (c.1500-1610), un brahmane de la famille Parakroḍa, était un contemporain de Nilākaṇṭha[33], qu'il qualifie d'Ācārya, maître[34]. La *Yuktibhāṣā* n'est pas, à proprement parler, un commentaire du *Tantrasaṅgraha*, mais « (...) son but est de fournir les connaissances de base nécessaires à celui qui veut étudier le calcul des mouvements planétaires comme les décrit le *Tantrasaṅgraha*. Elle atteint ce but en exposant les concepts et théories de base des mathématiques et de l'astronomie, en énonçant les définitions et en explicitant les méthodologies et leurs développement logiques ».[35]

L'établissement des tables selon la *Yuktibhāṣā*

1. Notions de base

Commençons par les notions de base du chapitre 7 de la *Yuktibhāṣā*.[36] On construit un cercle de circonférence 21600′ et deux de ses diamètres, nord-

krit Series n°188, Trivandrum, 1958.

[31] Plus de détails dans K.V.Sarma, *Tantrasaṅgraha...*, pp.xxiv-xxxvi.

[32] Aussi nommé *Gaṇitanyāyasaṇgraha*, selon K.V.Sarma, *A History of the Kerala School ...*, p.60. La *Yuktibhāṣā* originale est en malayalam, mais il existe une *Gaṇitayuktibhāṣā*, composée en sanskrit, selon K.V.Sarma (*Tantrasaṅgraha...*, p.xlvi). Comme nous n'avons pas eu accès à cette version sanskrite, nous avons utilisé la traduction anglaise, par K.V.Sarma, du texte malayalam, dans une version de travail qui nous avait été aimablement communiquée par G.G.Joseph. Depuis, nous avons eu accès au texte publié *Gaṇita-yukti-bhāṣā* (Rationales in Mathematical Astronomy) of Jyeṣṭhadeva, Malayalam Text critically Edited with English Translation by K.V.Sarma, With Explanatory Notes by K.Ramasubramanian, M.D.Srinivas, M.S.Sriram, New Delhi, 2008.

[33] K.V.Sarma, *A History of the Kerala School ...*, p.59.

[34] Les manuscripts conservés de cet ouvrage ne mentionnent pas le nom de l'auteur, mais K.V.Sarma réussit à l'identifier grâce à des sources externes (voir K.V.Sarma, « *Jyeṣṭhadeva and his identification as the author of Yuktibhāṣā* », *Adyar Library Bulletin*, 22 (1958), pp.35-40).

[35] K.V.Sarma, *Tantrasaṅgraha...*, p.xlv.

[36] Le texte qui suit est une paraphrase, parfois fortement résumée, parfois traduite en symboles modernes, d'une traduction anglaise, par K.V.Sarma, de la *Yuktibhāṣā*.

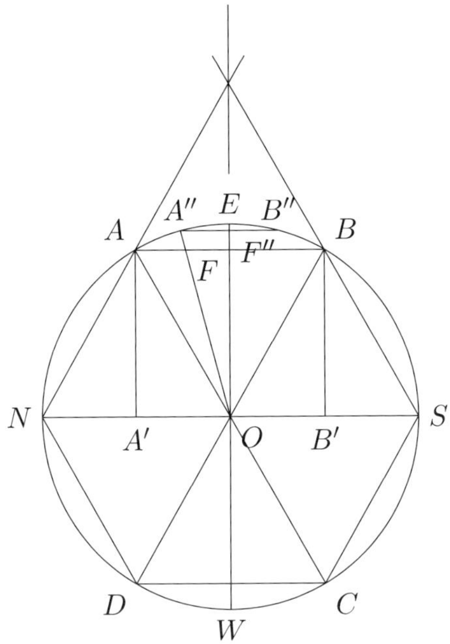

Figure 4

sud (NS 'horizontal') et est-ouest (EW 'vertical'). Deux triangles équilatéraux, NAO et OBS, sont tracés et leur symétrie est vérifiée à l'aide d'un fil à plomb, c'est-à-dire que les deux segments NA et SB sont prolongés au-delà de E de sorte qu'un fil à plomb (aussi nommé *lamba* 'perpendiculaire')[37], suspendu à leur intersection, passe par O. Comme les deux perpendiculaires au diamètre NS[38], par A et B, déterminent deux points A' et B' coupant chaque demi-diamètre en deux segments égaux, la distance entre A et B égale la longueur du demi-diamètre, égale aussi les longueurs des autres côtés des deux triangles équilatéraux.

En construisant deux triangles équilatéraux dans la partie ouest du cercle, on obtient un hexagone régulier inscrit $NABSCD$.

On définit ensuite le 'sinus complet' (samasta-jyā) de deux rāśi[39] (AB), égal au rayon du cercle. Un 'sinus complet', qui est une corde entière, ne

[37]Remarquer cette utilisation, en géométrie, d'un terme provenant de la pratique artisanale.

[38]Parallèles au 'fil à plomb', donc.

[39]En astronomie, un rāśi est un signe du zodiaque ou son étendue, donc $1800' = 30°$. Nous garderons cette dénomination dans la suite de cet article, pour rester plus près du texte et des raisonnements de la *Yuktibhāṣā*. Le rayon est parfois nommé trijyā 'sinus de trois [rāśi]'.

doit pas être confondu, dans le texte de la *Yuktibhāṣā*, avec le 'demi-sinus' (ardha-jyā), souvent nommé 'sinus' (jyā), qui est la demi-corde (AF pour l'arc d'un rāśi). Nous avons signalé plus haut que ces 'sinus' sont en fait des $rsinus$, avec $rsinus\ a = rsin(a/r)$, où a est la mesure d'un arc et a/r l'arc mesuré en radians.

Le 'sinus verse' (śara 'flèche') est ensuite défini comme la distance (EF pour l'arc d'un rāśi) du milieu du 'rsinus complet' au milieu de l'arc correspondant, et le 'rcosinus' comme la distance du milieu du 'rsinus complet' au centre du cercle (OF pour l'arc d'un rāśi).

La relation suivante lie l'ardha-jyā, le śara et le samasta-jyā (la corde) d'un rāśi :

$([rsinus(1\text{rāśi})]^2 + [\text{śara}(1\text{rāśi})]^2)^{1/2} = \sqrt{AF^2 + FE^2} = AE =$ samasta-jyā (1 rāśi).

En déplaçant maintenant la corde AE pour qu'elle soit perpendiculaire au diamètre est-ouest, on observe que sa moitié $A''F'$, qui est l'ardha- jyā d'$1/2$ rāśi, permet de trouver le $rcosinus$ d'$1/2$ rāśi par une relation similaire :

$rcosinus(1/2\text{rāśi}) = OF' = \sqrt{(A''O)^2 - (AE/2)^2} = rsinus(2,5\text{rāśi})$

et

$\text{śara}(1/2\text{rāśi}) = OE - OF' = r - rcosinus(1/2\text{rāśi})$.

L'auteur note que cette méthode permet de déduire les $rsinus$ des (demi-)arcs successifs à l'aide de carrés et de racines carrées seulement. Cela rappelle le Pauliśasiddhānta (IV, 4-5), l'énoncé de *Āryabhaṭīya*, II.11 et son commentaire par Bhāskara I.

2. Construction des formules itératives de base *(voir Figure 5)*

Le cercle est maintenant divisé en quatre quadrants (pāda), chacun d'entre eux divisé en 24 'portions d'arc' d'$1/8$ rāśi (correspondant à l'angle $225'$) par 25 points ($A_0 = E$, A_1, A_2, etc.) sur le cercle.

A partir de chaque point, le segment horizontal menant au diamètre est-ouest définit le rsinus de l'arc correspondant et le segment vertical menant au diamètre nord-sud définit le rcosinus (par exemple, $B_1,A_1 = rsinus(1/8\text{rāśi}) = rsinus\ n°1$). Les diamètres est-ouest et nord-sud sont les bases des rsinus et rcosinus respectivement, et leurs sommets sont les points de division.

Chaque 'portion d'arc' a aussi sa base et son sommet mais, pour des raisons pratiques, elles sont définies différemment selon qu'on envisage les $rsinus$ ou les $rcosinus$. La 'rsinus-base' d'une 'portion d'arc' est son extrémité la plus proche du diamètre est-ouest, tandis que son 'rsinus-sommet' est son extrémité la plus éloignée. C'est le contraire pour le $rcosinus$.

Tous les segments ainsi définis, en vue de visualiser les $rsinus$ et $rcosinus$, découpent sur les autres des sous-segments dont les plus proches du cercle sont nommés bhuja-khaṇḍa ('rsinus partiels') et koṭi-khaṇḍa ('rcosinus partiels'). Par exemple, le premier bhuja-khaṇḍa est le $rsinus$ de la première 'portion d'arc', soit $rsinus$ n°1 (B_1,A_1), le second bhuja-khaṇḍa est la portion du second $rsinus$ allant du sommet de celui-ci au $rcosinus$ de la première 'portion d'arc', soit H_3A_2, etc.

Notons que ces 'rsinus et rcosinus partiels' forment les côtés de l'angle droit de triangles rectangles dont les hypoténuses sont égales à la corde de la 'portion d'arc' commune. Nous avons donc 24 triangles rectangles de même hypoténuse et de côtés différents.

Un arc quelconque, allant du point E à un point du premier quadrant, est nommé iṣṭa-bhuja-cāpa ('arc dont le $rsinus$ est autant'[40]). Si cet arc est issu de N, il est nommé iṣṭa-koṭi-cāpa ('arc dont le $rcosinus$ est autant').[41]

On peut calculer les côtés de ces petits triangles ayant la même hypoténuse (la corde de $225'$), car ce sont des 'rsinus et rcosinus partiels', et les tabuler (paṭhita-jyā) comme dans d'autres textes.[42]

Si un point se trouve à l'extrémité d'un arc (multiple de $225'$), son $rsinus$ sera dans la table. Sinon, il faut ajouter la partie proportionnelle de l'augmentation par rapport à l'extrémité tabulée. Mais une méthode simple, comme la règle de trois, ne fonctionne pas, car si le deuxième arc égale 2 fois le premier, le troisième 3 fois le premier, cela n'est pas vrai des $rsinus$. En fait, cela sera vrai des premiers arcs, presque plats parce que les śara sont infimes, mais pas des suivants, plus courbés quand le śara augmente.[43]

Pour calculer une table précise de $rsinus$, on trace la 'portion d'arc' de $225'$ symétriquement par rapport à la ligne EW ($F_1E = 112,5'$) de sorte que $F_1G_1 = rsinus\ n°0,5 = rsinus 112,5' \approx 112,5'$, et le même arc à partir de E ($A_1E = 225'$). On a $rcosinus\ n°0,5 = rcosinus 112,5' = \sqrt{r^2 - 112,5^2} = rsinus\ n°23,5$ et $r - rsinus\ n°23,5 = śara\ \ n°0,5$.

En comparant les triangles semblables, Jyeṣṭhadeva obtient d'autres $rsinus$ et $rcosinus$ tabulés.

La similitude entre les triangles A_1B_1E et OG_1F_1 fournit :

[40] Autant traduit iṣṭa 'désiré', c'est-à-dire 'autant qu'on désire'.

[41] Ces expressions sont donc comparables à nos $arcsin$ (ou sin^{-1}) et $arccos$ (ou cos^{-1}).

[42] Malheureusement, les titres ou auteurs de ces textes ne sont pas mentionnés par Jyeṣṭhadeva.

[43] Intéressante remarque. Nous verrons plus loin que Jyeṣṭhadeva propose une meilleure méthode d'interpolation.

$rsin\ n°1 = A_1B_1 = (OG_1/OF_1) \times A_1E = ([rcos\ n°0,5]/r) \times 225 = ([rsin\ n°23,5]/r) \times 225$, et
$rcos\ n°1 = rsin\ n°23 = A_1C_1 = EO - EB_1 = EO - (F_1G_1/OF_1) \times A_1E = r - ([rsin\ n°0,5]/r) \times A_1E = r - ([rsin\ n°0,5]/r) \times 225$

La similitude entre les triangles $F_2H_2F_1$ et OB_1A_1 fournit :
$rsin\ n°1,5 = F_2G_2 = H_2G_2 + F_2H_2 = H_2G_2 + (A_1C_1/A_1O) \times F_2F_1 = rsin\ n°0,5 + ([rcos\ n°1]/r) \times F_2F_1 = rsin\ n°0,5 + ([rsin\ n°23]/r) \times 225$, et
$rcos\ n°1,5 = rsin\ n°22,5 = F_2D_2 = F_1D_1 - F_1H_2 = F_1D_1 - (A_1B_1/A_1O) \times F_2F_1 = rcos\ n°0,5 - ([rsin\ n°1]/r) \times F_2F_1 = rcos\ n°0,5 - ([rsin\ n°1]/r) \times 225 = rsin\ n°23,5 - ([rsin\ n°1]/r) \times 225$

La similitude entre les triangles $A_2H_3A_1$ et OG_2F_2 fournit :
$rsin\ n°2 = A_2B_2 = B_2H_3 + H_3A_2 = B_2H_3 + (G_2O/OF_2) \times A_2A_1 = rsin\ n°1 + ([rcos\ n°1,5]/r) \times A_2A_1 = rsin\ n°1 + ([rsin\ n°22,5]/r) \times 225$, et
$rcos\ n°2 = rsin\ n°22 = A_2C_2 = A1C1 - A_1H_3 = A_1C_1 - (F_2G_2/OF_2) \times A_2A_1 = rcos\ n°1 - ([rsin\ n°1,5]/r) \times A_2A_1 = rcos\ n°1 - ([rsin\ n°1,5]/r) \times 225 = rsin\ n°23 - ([rsin\ n°1,5]/r) \times 225$

On peut donc, en commençant par $rsin\ n°0,5$ et $rcos\ n°0,5 = rsin\ n°23,5$, tabuler tous les $rsin\ n°i$ et $n°(i+0,5)$ ($i = 1$ à 23) et ensuite rejeter les $n°(i+0,5)$ si nécessaire.

En symboles modernes et en remplaçant $n°i$ par $N \times 112,5'$ (avec $N = 2i$), les formules itératives sont les suivantes :

$$rsin(N \times 112,5') = rsin([N-2] \times 112,5') + [rcos([N-1] \times 112,5') \times 225]/r \quad (1)$$

et

$$rcos(N \times 112,5') = rcos([N-2] \times 112,5') - [rsin([N-1] \times 112,5') \times 225]/r \quad (2)$$

où N est un entier compris entre 2 et 48, et $r \approx 3437,75$.

Les premières valeurs (et leurs équivalents modernes) obtenues par ce procédé sont données dans la table 3 :

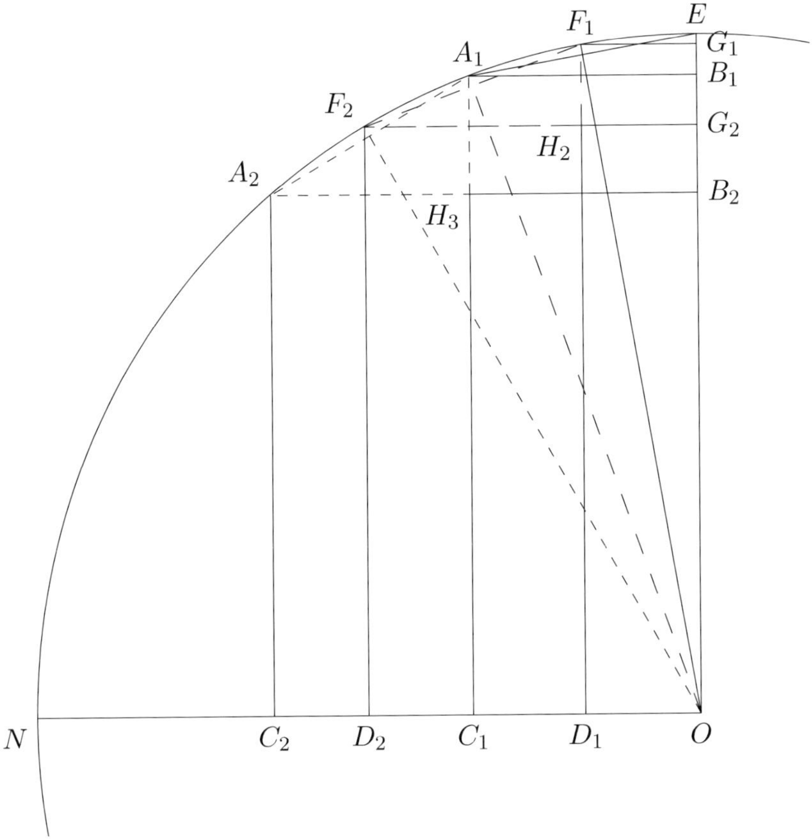

Figure 5

Table 3

N	angle correspondant à l'arc (N×112.5')	rsin (N×112.5')	sinus (moderne)	rcos (N×112.5')
2	3°45'	224.88	0.065415 (0.065403)	3430.39
3	5°37.5'	337.02	0.098035 (0.098017)	3421.03
4	7°30'	448.79	0.130548 (0.130526)	3408.33
5	9°22.5'	560.09	0.162923 (0.162895)	3391.66
6	11°15'	670.77	0.195119 (0.195090)	3371.67

3. La formule itérative d'Āryabhaṭa explicitée

Dans la section suivante, Jyeṣṭhadeva cherche une relation entre le $rsinus$ d'une certaine 'portion d'arc' et la différence seconde des $rsinus$. En notations modernes, le texte équivaut à :
$[rcos([N-1] \times 112,5') - rcos([N+1] \times 112,5')] \times \frac{225}{r} =$
$[rsin(N \times 112,5') - rsin([N-2] \times 112,5')] - [rsin([N+2] \times 112,5') - rsin(N \times 112,5')]$,
une identité qui résulte immédiatement de la relation (1).

Et, puisque (2) $\Rightarrow rcos([N-1] \times 112,5') - rcos([N+1] \times 112,5') = rsin(N \times 112,5') \times \frac{225}{r}$, on a :
$rsin(N \times 112,5') \times (\frac{225}{r})^2 =$
$[rsin(N \times 112,5') - rsin([N-2] \times 112,5')] - [rsin([N+2] \times 112,5') - rsin(N \times 112,5')]$,
ce que le texte exprime : « *Quand le premier sinus est multiplié par le carré du sinus complet de la portion d'arc et divisé par le carré du trijyā,*[44] *le résultat sera la différence entre le premier sinus partiel et le second sinus partiel* ».[45]

Avec $r = 3438$, on trouve bien $(\frac{225}{r})^2 \approx 233,4784$, et avec $r = \frac{21600}{2\pi}$, $(\frac{225}{r})^2 \approx 233,444$.

Quelques méthodes d'interpolation

Bien qu'il soit avéré que Jyeṣṭhadeva, et peut-être déjà Nilākaṇṭha ou même Mādhava (c.1340-1425), connaissait les séries équivalentes à celles de Taylor pour les fonctions $sinus$, $cosinus$ et $arc\ tangente$[46], nous nous limiterons ici à des méthodes d'interpolation du $sinus$ ne dépassant pas le second degré[47].

Ci-dessus, nous avons vu Jyeṣṭhadeva faire allusion à la question du calcul du $rsinus$ d'un arc non tabulé. Dans une autre section du chapitre, il donne la méthode suivante (en notations modernes) :

Le $rsinus$ de l'arc EP doit être calculé. D'abord, on prend le point tabulé A_N le plus proche de P. L'arc $PA_N = a$ est le śiṣṭa-cāpa ('arc restant').

---

[44] Ou tri-rāśi-jyā, '$rsinus$ de trois signes du zodiaque', soit $rsinus 5400'$ (arc équivalent à $90°) = r$.

[45] Pour plus de détails, voir R.C.Gupta, « *Early Indians on second order sine differences* », *Indian Journal of History of Science*, 7 (1972), pp.81-86.

[46] Selon D.Bressoud, « *Was Calculus invented in India ?* », *The College Mathematics Journal*, 33-1 (January 2002), pp.1-13. Note de l'éditeur : Notons que la série pour $arc\ tangente$

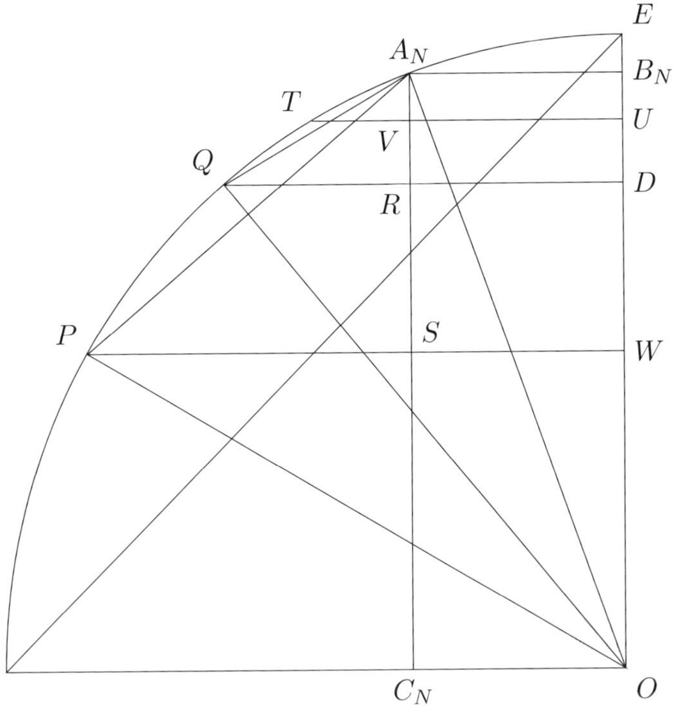

Figure 6

Grâce aux triangles semblables PA_NS et OQD (figure 6), on a : $PA_N \times QD/OQ = A_NS$ et $PA_N \times DO/OQ = PS$ (3).

Comme QD et DO sont inconnus, la méthode suivante est indiquée : on prend la corde QA_N comme hypoténuse et ses $rsinus$ et $rcosinus$ comme « icchāphala à appliquer aux $rsinus$ tabulaires correspondants ».

Cela signifie calculer $QD = DR + RQ = A_NB_N + RQetDO = OB_N - DB_N = A_NC_N - RA_N$, puis assimiler $QD \approx A_NB_N$ et $DO \approx A_NC_N$ (puisque RQ et RA_N sont inconnus, mais négligeables car $< 112,5'/2$) dans (3) : $PA_N \times A_NB_N/OQ = A_NS$ et $PA_N \times A_NC_N/OQ = PS$.

En termes de $rsinus$: $rsin(N \times 112,5' + a) = PW = SW + PS = A_NB_N + PA_N \times QD/OQ = A_NB_N + PA_N \times A_NC_N/r$
$\Leftrightarrow rsin(N \times 112,5' + a) = rsin(N \times 112,5') + \frac{a \times rcos(N \times 112,5')}{r}$

permet de calculer π avec une précision arbitraire, comme le montrent K.Mahesh et al., p.189.

[47] Selon D.Bressoud, *op.cit.*, p.8, Brahmagupta aurait déjà donné une méthode d'interpolation équivalente.

Jyeṣṭhadeva conclut cette section en disant que, si l'on désire plus de précision, on peut prendre un quart du śiṣṭa-cāpa, ou même la moitié (du quart ?), comme corde complète, et il cite partiellement un vers[48] attribué à l'astronome Mādhava par K.V.Sarma. Ce vers n'est pas très clair, mais on peut probablement comprendre que, pour calculer des valeurs plus précises de QD et OD, on égale TU et UO (voir figure 6) à A_NB_N et A_NC_N respectivement. En remplaçant dans les équations (3), on obtient : $rsin(N \times 112,5' + a) = A_NB_N + PA_N \times DO/OQ$, où DO, au lieu d'être égalé à A_NC_N, est remplacé par $A_NC_N - RA_N$, avec $RA_N = QA_N \times TU/OT$ (puisque RQA_N et UOT sont des triangles semblables).

Finalement, $rsin(N \times 112,5' + a) =$
$A_NB_N + PA_N \times (A_NC_N - RA_N)/OQ$
$= A_NB_N + PA_N \times (A_NC_N - QA_N \times TU/OT)/OQ$
$\approx A_NB_N + PA_N \times (A_NC_N - QA_N \times A_NB_N/OT)/OQ,$

ou encore $rsin(N \times 112,5' + a)$
$= rsin(N \times 112,5') + \frac{a}{r} \times [r\cos(N \times 112,5') - \frac{a}{2} \cdot \frac{r\sin(N\times 112,5')}{r}]$

Cette expression est équivalente aux trois premiers termes du développement de Taylor :

$\sin(N \times 112,5' + \alpha)$
$= sin(N \times 112,5') + \alpha \times \cos(N \times 112,5') - \frac{1}{2}\alpha^2 \times \sin(N \times 112,5'),$
où $\alpha = \frac{a}{r}$

En disant que le procédé peut être poussé plus loin, Jyeṣṭhadeva semble indiquer qu'il pourrait calculer les termes suivants de ce développement.

On retrouve le vers auquel Jyeṣṭhadeva faisait allusion dans le *Tantrasaṅgraha*, chapitre 2 :[49]

iṣṭadoḥkoṭidhanuṣoḥ svasamīpa samīrite //10//
jye dve sāvayave nyasya[50] kuryādūnādhikaṃ dhanuḥ /
dvighnatalliptikāptā eka-śara-śaila-śikhi-indavaḥ //11//
nyasyācchedāya ca mithastatsaṁskāravidhitsayā /
chitvaikāṃ prāk kṣipejjahyāt taddhanuṣyadhikonake //12//
anyasyāmatha tāṁ dvighnāṁ tathā 'syāmity[51] saṁskṛtiḥ /
iti te kṛtasaṁskāre svaguṇau dhanuṣostayoḥ //13//

[48]Commençant par iṣṭadoḥkoṭidhanuṣoḥ.
[49]K.V.Sarma, *Tantrasaṅgraha*..., p.112, et S.K.Pillai, *The Tantrasaṅgraha*, ..., p.19.
[50]'nyasya pour anyasya dans l'édition de Trivandrum.
[51]syām dans l'édition de Trivandrum.

« *Après avoir placé deux cordes, faites de parts*[52], *près des (arcs)*[53] *rcosinus [et rsinus] cherchés, calculez l'arc additionnel ou manquant. Prendre comme dénominateur un-cinq-sept-trois-un (13571) divisé par les minutes*[54] *de cet [arc additionnel ou manquant] multipliés par deux, dans le but d'améliorer ces [rsinus et rcosinus cherchés] l'un par rapport à l'autre. Après avoir divisé un [rcosinus*[55]*], ajouter ou soustraire [le résultat] au précédent [rsinus], selon le [caractère] additionnel ou manquant de l'arc. La méthode est la même pour ce [dernier résultat] doublé [et ajouté ou soustrait] à l'autre*[56]. *Deux améliorations appropriées ont été ainsi apportées aux deux (arcs)* ».

Pour traduire la première moitié du vers 13, nous avons eu recours au commentaire :

athaivaṁ kṛtāṁ tāṁ dviguṇitāṁ kṛtvā pūrvoktenaiva hārakeṇa vibhajya labdhaṁ yat phalaṁ tat punaranyasyāṁ sādhyajyāyāṁ eva taṁ dhanuṣa ūnādhikavaśād ṛṇaṁ dhanaṁ vā kuryāt /

« *Après avoir doublé cette [quantité juste] obtenue, [l']avoir divisée par le diviseur, le résultat, qui est un quotient, qu'on l'ajoute encore à l'autre corde comme une quantité positive ou négative selon le caractère additionnel ou manquant de l'arc* »

Il s'agit d'un extrait du *Laghuvivṛti*, 'commentaire facile' au *Tantrasaṅgraha*, composé par Śaṇkara Vāriyar (c.1500-1560), qui est aussi l'auteur de la *Yuktidīpikā*, un autre commentaire au *Tantrasaṅgraha*, et d'une *Kriyākramakarī*, un commentaire inachevé[57] sur la *Līlāvatī* de Bhāskara II.

[52]Cette précision, sāvayave, assure qu'il s'agit de 'portions d'arc', multiples du même arc de base.

[53]Le duel dhanuṣoḥ montre ici qu'il s'agit des deux 'cordes' sinus et cosinus, de sorte que arcs serait une traduction inappropriée. On pourrait aussi traduire 'proche de l'arc, dont le sinus et le cosinus sont recherchés' en remplaçant le duel par un singulier.

[54]Liptikā = liptā 'minute d'arc', un mot d'origine grecque.

[55]Il faut choisir l'un des jyā, l'autre sera traité par le vers suivant.

[56]rsinus ou rcosinus, selon ce que l'on aura choisi dans l'opération précédente (vers précédent).

[57]Par Śaṇkara, mais il a été terminé par Nārāyaṇa, quand il avait seulement dix-huit ans. Il commence sa partie du texte par les mots « *Ce commentaire a été composé par Tṛkkuṭaveli Śaṇkara Vāriyar. J'ai entendu dire par Paraṇṇoṭṭu qu'il avait été composé avec grand soin à la demande d'Āzhāñceri.* » Cela nous assure que le premier auteur était bien Śaṅkara et que Nārāyaṇa (c.1540-1610) était un contemporain plus jeune de Jyeṣṭhadeva, mentionné ici par son nom de famille Paraṇṇoṭṭu. L'Āzhāñceri mentionné est Nārāyaṇa Āzhāñceri Tāmprakkal, chef religieux des brāhmanes nampūtiri, auquel Nīlakaṇṭha se réfère comme son patron (*Līlāvatīof Bhāskarācārya with Kriyākramakarī of Śaṅkara and Nārāyaṇa, critically edited by K.V.Sarma*, Vishveshvaranand Institute, Hoshiarpur, 1975, p.xvii).

Śaṅkara était aussi l'élève de Nīlakaṇṭha.[58]

Les opérations sont les suivantes :

En supposant les rsinus et rcosinus de $N \times 112,5'$ tabulés, nous voulons calculer

$rsin([N \times 112,5'] + a)$, avec $0 < a < 112,5'$

D'abord, on calcule $\frac{13751}{2\times a} = \frac{4\times 3437.75}{2\times a} = \frac{2\times 3437.75}{a} = \frac{2\times r}{a}$,

On divise $rsin(N \times 112,5')$ par ce nombre et on soustrait le résultat de $rcos(N \times 112,5')$:

$rcos(N \times 112,5') - [rsin(N \times 112,5')]/\frac{2\times r}{a} =$

$rcos(N \times 112,5') - \frac{a}{2\times r} \times rsin(N \times 112,5') =$

$r \times [cos(N \times 112,5') - \frac{1}{2} \times [sin(N \times 112,5')] \times \alpha]$, où $\alpha = \frac{a}{r}$

Ensuite, on divise ce résultat, doublé, par $\frac{2\times r}{a}$, et on l'ajoute à $rsin(N \times 112,5')$:

$rsin(N \times 112,5') + 2[rcos(N \times 112,5') - [rsin(N \times 112,5')]/\frac{2\times r}{a}]/\frac{2\times r}{a} =$

$rsin(N\times 112,5') + 2\times r\times [cos(N\times 112,5') - \frac{1}{2}\times [sin(N\times 112,5')]\times\alpha]\times\frac{\alpha}{2} =$

$r\times [sin(N\times 112.5') + cos(N\times 112.5')\times\alpha - \frac{1}{2}\times [sin(N\times 112.5')]\times\alpha^2]$,

qui contient les trois premiers termes du développement de Taylor de $sin([N \times 112,5'] + \alpha)$, comme annoncé.

Le *Tantrasaṅgraha* donne donc :

$rsin(N \times 112.5' + a)$

$= r\times [sin(N\times 112.5') + cos(N\times 112.5')\times\alpha - \frac{1}{2}\times [sin(N\times 112.5')]\times\alpha^2]$.

Nous avons vu aussi que la *Yuktibhāṣā* proposait de continuer le processus en divisant l'arc par 2, 4, etc.

On trouve un écho de ce processus, rapidement abandonné, dans le *Laghuvivṛti* :

yadapyatra tataḥ pūrvaṁ koṭijyārdhataḥ tenaiva hāreṇa labdhaṁ dorjyāyāḥ tataḥ pūrvaṁ dorjyācaturaṁśāt koṭijyāyāṁ tataḥ pūrvaṁ koṭijyāyāṣṭamāṁśato dorjyāyāṁ tataḥ ṣodaśāṁśataḥ koṭijyāyāṁ ca labdhaphalaṁ kartavyameva / tathāpi tasyālpatvādevopekṣitamiti mantavyam /

« *Même si, dans ce cas, le sinus est obtenu à partir du demi-cosinus par le diviseur, le résultat obtenu à partir du quart du sinus doit être placé dans le cosinus, ensuite [le résultat obtenu] à partir du huitième du cosinus [doit être placé] dans le sinus et ensuite [le résultat obtenu] à partir du seizième [du sinus doit être placé] dans le cosinus. Cependant, il faut comprendre que [cela est] ignoré du fait de son caractère infime.* »

[58]K.V.Sarma, *Tantrasaṅgraha...*, p.lxiii.

Jusqu'ici, le développement obtenu correspondait bien au développement de Taylor, mais la suggestion d'utiliser les quotients de l'arc par les puissances suivantes de deux amènerait de nouveaux termes différents des termes correspondants du développement de Taylor. Nous suivrons donc la suggestion de Śaṅkara et en resterons aux termes de degré ≤ 2 dans cette explication géométrique des découvertes kéralaises dans le domaine de la trigonométrie.

Mathématiques et scolastique dans l'Inde médiévale, l'exemple du Haricarita de Parameśvara Bhaṭṭa

Pierre-Sylvain Filliozat
Membre de l'Institut

Les hommes qui dans l'Inde antique et médiévale ont fait œuvre de science et exposé leur savoir, ont été avant tout des lettrés à la manière indienne. Ceci implique qu'ils ont reçu une formation scolastique qui va bien plus loin que l'apprentissage de la langue sanskrite, parce qu'elle comprend une réflexion approfondie sur la nature du langage et une technique d'utilisation et d'interprétation. Tout savant, tout auteur sanskrit part de cette culture et pour accéder à leurs œuvres, il faut aussi passer par leur scolastique. Je propose ici de montrer comment des astronomes du Kerala ont réalisé une table astronomique de forme tout à fait originale et comment elle a pu être agrémentée dans une enveloppe poétique et religieuse encore plus admirable.

Le *Haricarita*[1] se présente par son titre « La geste de Hari » comme une composition en vers sur un sujet de mythologie viṣṇuite. Et c'est bien ce qu'il est, dans une première lecture : un résumé du Chant X du *Bhāgavata Purāṇa* consacré à l'enfance de Kṛṣṇa. Dans une seconde lecture on reconnait au début de chaque strophe les *vākya* de Vararuci qui sont une table astronomique servant à calculer les longitudes de la lune. Il s'agit donc d'un exercice de virtuosité littéraire : combiner une information astronomique avec une composition de poésie dévotionnelle. Pour rendre justice à ce texte il nous faut présenter son contenu astronomique, sa méthode de composition, sa valeur poétique et son inspiration religieuse.

[1]*Haricarita by Parameśvara Bhaṭṭa (of Payyūr)*, edited by V. Krishnamacharya, The Adyar Library Series–No. 63, The Adyar Library, Madras, 1948, pages l & 88. Cette édition est accompagnée d'un commentaire sanscrit de son éditeur, d'une préface et d'une longue introduction de C. Kunhan Raja.

Les *vākya* « paroles » sont de courtes propositions qui ne sont pas toujours des phrases complètes, mais qui ont toujours quelque sens. Faites de quatre à six syllabes, elles sont en fait des énoncés de nombres, sur le principe du code *kaṭapayādi*. Ce code bien connu est un procédé mnémotechnique reposant sur le fait que la mémoire s'attache plus aisément à des mots qu'à des nombres. Ce n'est donc pas le sens des mots choisis qui importe, c'est le nombre qu'ils notent. Mais les mots pourvus d'un sens quelconque s'impriment dans la mémoire et peuvent toujours être décodés. Ce code a été conçu à date ancienne dans le sud de l'Inde. Il a été très utilisé au Tamilnāḍu et au Kerala.

Kaṭapayādi signifie littéralement « ce qui commence par *k*, *ṭ*, *p* ou *y* ». Ce procédé note les chiffres de 1 à 9 par des lettres de l'alphabet, en commençant par l'une ou l'autre de ces quatre lettres. Les voyelles indépendantes d'une consonne, les lettres *ñ* et *n* notent 0. Dans un groupe de consonnes c'est la dernière du groupe qui a une signification numérique. La première syllabe du mot note l'unité, la seconde la dizaine et ainsi de suite. Le *vākya* qui se lit de gauche à droite se transcrit donc en chiffres de droite à gauche.

1	2	3	4	5	6	7	8	9	0
k	kh	g	gh	ṅ	c	ch	j	jh	ñ
ṭ	ṭh	ḍ	ḍh	ṇ	t	th	d	dh	n
p	ph	b	bh	m					
y	r	l	v	ś	ṣ	s	h	ḷ	

Vararuci, à propos de qui l'on ne connait qu'une légende, aurait, en utilisant ce code, composé une table servant à calculer la longitude de la lune pour un jour donné, par exemple :

1. *gīr naḥ śreyaḥ* « puisse [la déesse] Parole nous [donner] le souverain bien », ce qui donne les nombres [00]-12-03.

. . .

26. *dhanyā vidyeyaṁ syāt* « que bénéfique soit cette connaissance », ce qui donne les nombres 11-14-19.

. . .

La table comprend ainsi 248 *vākya*. Chacun note une série de trois nombres. Ce sont successivement un des douze *rāśi* « signes du zodiaque », un des 30 *tithi* qui partagent le *rāśi*, un des 60 *ili* (/*nāḍī*) ou parties du *tithi*.

L'usage pratique de cette table est décrit en détails dans plusieurs textes d'astronomie du Kerala. Le plus utilisé de tous est le *Vākyakaraṇa* anonyme, commenté entre autres par Sundararāja[2]. Sa date a été ingénieusement déterminée par ses éditeurs comme située entre 1282 et 1306. Cette table astronomique était apprise par cœur dès le plus jeune âge dans les familles d'astrologues et dans toute famille orthodoxe gardienne de la tradition savante. Elle permettait de calculer mentalement la position quotidienne de la lune, le *tithi* et le *nakṣatra* correspondants, à l'aide des règles d'un manuel comme le *Vākyakaraṇa* lui aussi mémorisé. Cela était considéré comme les rudiments de base de la science. C'est ainsi que le poète du Kerala Mepputtūr Nārāyaṇa Bhaṭṭa se lamente sur la mort d'un grand savant, Tṛkkaṇṭiyūr Acyuta Piṣāraka (/Piṣāroṭi), son maître, en disant qu'après lui on ne saura rien de plus que calculer *tithi* et *nakṣatra* :

hanta jyotiṣatantra paryavasitā tithyṛkṣayos te kathā
dhārṣṭyaikapravaṇāsi vaidyasaraṇe naṣṭo'sy alaṁkāra hā |
he śabdāgama nirdayaṁ vibudhatālubdhair nipīḍiṣyase
vidyātmā svar asarpad adya bhavatām ādhārabhūr acyutaḥ ||

« Ô science des luminaires célestes, hélas ! ton histoire se termine dans [le calcul] de jours lunaires et d'étoiles. Tu es vouée à la seule témérité, voie du médecin. Tu es ruinée, ô poétique. Ô science de la parole, sans pitié tu seras pilonnée par des cuistres. Un être à l'âme de science s'en est allé au paradis aujourd'hui, celui qui était votre support, Acyuta. »

Ces calculs furent aussi les premières informations que l'Europe recueillit sur l'astronomie indienne. Nous citerons l'exposé de l'astronome Legentil qui au cours d'un séjour dans le sud de l'Inde entre 1761 et 1769 s'est enquis des pratiques astronomiques de cette région directement auprès de savants tamouls à Pondichéry. Il en nomme deux, Nana Moutou et Maridas. C'est sans doute auprès du second, lettré tamoul qui avait reçu une formation des jésuites en latin et français, qu'il a acquis les meilleures connaissances. Dans sa belle relation de voyage, publiée en 1779, intitulée *Voyage dans les mers de l'Inde fait par ordre du roi, à l'occasion du passage de Vénus sur le disque du soleil le 6 juin 1761 et le 3 du même mois 1769*, Legentil expose ce qu'il a appris et donne, à défaut des *vākya* sanskrits, la table de leurs valeurs numériques. Maridas, quant à lui, a composé en français un manuel très détaillé, intitulé *Mémoire sur l'astronomie en vigueur sur la côte de Coromandel, envoyé à M. de Monneron, en 1787, par Maridas-poullé, interprète du Roi, à Pondichéry*. Le manuscrit envoyé par son auteur se trouve aujourd'hui à

---

[2] *Vākyakaraṇa with the commentary Laghuprakāśikā by Sundararāja*, critically edited with Introduction, Appendices etc. by T. S. Kuppanna Sastri and K. V. Sarma, Kuppuswami Sastri Research Institute, Madras, 1962.

la Bibliothèque nationale de France à Paris[3]. M. François Patte en prépare l'édition. En cette même année 1787 paraissait à Paris, chez Debure l'Aîné, le *Traité de l'astronomie indienne et orientale* de Jean-Sylvain Bailly, qui devait s'illustrer au Jeu de Paume en 1789. Bailly n'a pas eu accès direct aux sources indiennes, mais s'est appuyé sur les informations envoyées en France par des missionnaires et par Legentil.

Pour calculer la longitude de la lune à une date donnée, on part du *śuddhadina* qui est le nombre de jours complets depuis le début du Kaliyuga. T. S. Kuppanna Sastri et K. V. Sarma, éditeurs du *Vākyakaraṇa* ont fait le calcul pour l'année Śaka écoulée 1869 et la fin du 24 du mois de Kanyā (27 septembre 1948) pour obtenir 1844004 jours de Kali, en suivant la règle donnée au début de leur texte :

dhūsīkāla(3179)yutaḥ śākaḥ kalyabda iti kīrtitaḥ |
kalyabdo mātula(365)guṇaḥ varṣavāṁ(4)śena saṁyutaḥ |
punar abdān māna(5)guṇāt sālapriya(1237)vivarjitāt || 1.2 ||
tatsamā(576)ptair dinair yuktaṁ śukravārādikaṁ dinam |
sphuṭārkacakrāvadhikam

« L'année Śaka augmentée de 3179 est réputée être l'année de Kali. L'année de Kali multipliée par 365, additionnée du quart des années, de nouveau additionnée des jours obtenus à partir de l'année de Kali multipliée par 5, diminuée de 1237 et divisée par 576, est le nombre de jours jusqu'à la fin de l'année solaire vraie, le premier jour de Kali étant un vendredi. »

A partir de ce nombre on détermine le *vākya*. La formule suivante du *Vākyakaraṇa* en indique la méthode :

vededdhajñānatuṣṭo(1600984)nadivasād rasagairikaiḥ(12372) |
kālānalai(3031)r devarai(248)ś ca śiṣṭaṁ vākyaṁ śaśāṅkajam || 1.9 ||
gīrnaḥśreyādikaṁ jñeyaṁ kramāl labdhaphalais tribhiḥ |
nityadevasukhādhīnaṁ(09-27-48-10) yajñayogasunāyakam(11-07-31-01) || 1.10 ||
tanurviśvasukham(27-44-06) hatvā senānārināthanam(07-02-00-7) |
tadyoge nikṣipecchuddhaṁ vikalādyaṁ dhruvaṁ vidhoḥ || 1.11 ||

« Le reste du nombre de jours de Kali diminué de 1600984, divisé par 12372, 3031, 248 est connu comme un *vākya* de la lune dans la table *gīr naḥ śreyaḥ*... Après avoir multiplié par ces trois quotients successivement 09-27-48-10, 11-07-31-01, 00-27-44-06, on doit ajouter [à leur somme]

[3]Nouvelles acquisitions françaises 2626.

07-02-00-07 en addition avec le dit [*vākya*]. C'est la longitude vraie de la lune sans correction. »

Pour l'exemple cité de 1844004 jours de Kali, il y aura les opérations suivantes :

1844004 - 1600984 = 243020
243020 : 12372 = 19, reste 7952
7952 : 3031 = 2, reste 1890
1890 : 248 = 7, reste 154
Le *vākya* 154 est "jitvāsya rathaḥ" = 07-21-48
puis :
19 x 09-27-48-10 = 08-18-15-10
2 x 11-07-31-01 = 10-15-02-02
7 x 00-27-44-06 = 06-14-08-42
\+ 07-02-00-07
en additionnant les quatre résultats on obtient le *dhruva* [longitude]
soit 09-19-26-01
et en ajoutant le
vākya 154 + 07-21-48
on obtient le
vikalādidhruva = 04-11-14-01

Legentil présente les quatre diviseurs susdits comme « périodes lunaires des Brames, qui leur servent de diviseur » et les quatre multiplicateurs comme « périodes lunaires des Brames qui leur servent de multiplicateur ».

Des noms tamouls sont donnés à ces périodes :

vedam	1600984^{j}	+ 7^{mois}-2^{jours}-0^{heure}-7'
rassam	12372	+ 9-27-48-10
calam	3031	+ 11-7-31-1
devaram	248	+ 0-27-44-6

La période de 248 jours est utilisée depuis l'antiquité babylonienne. La lune parcourt en moyenne 13° de longitude en un jour, mais avec des fluctuations allant jusqu'à 6° en deçà ou au-delà de sa moyenne. On définit un mois anomalistique de la lune comme l'intervalle qui sépare deux passages consécutifs de la lune au périgée, durée de 27 j 13 h 18' 33". Le nombre 248 est un entier correspondant approximativement à 9 mois anomalistiques, 3031 à 110, 12372 à 449. Le calcul part du nombre de jours à partir du début de l'âge Kali, ce qui amène à l'utilisation d'un nombre très élevé. On en retranche donc une période appelée *śodhya* « à soustraire » [4]. Il reste un

[4] L'on a évidemment intérêt à prendre le *śodhya* le plus élevé. Le *Vākyakaraṇa* utilise le

nombre élevé de jours, dans lequel on peut trouver plusieurs de ces périodes de mois anomalistiques. On divise donc ce nombre de jours, d'abord par la plus grande période, puis le reste par la période suivante, finalement le dernier reste par la période de 248 jours qui est couverte par la liste des *vākya* qui suit l'ordre des jours. Le reste de la division par 248 sera le nombre de jours restants et le n° d'ordre d'un *vākya* dans la liste. On applique ensuite la règle d'utilisation du *vākya* ainsi obtenu.

Il y a une longue histoire des méthodes de calcul sur la base du nombre 248 pour le calcul de la longitude quotidienne de la lune. En Inde elle commence avec la *Pañcasiddhāntikā* de Varāhamihira. La méthode ci-dessus a été courante au Kerala et au Tamilnāḍu.[5]

Le texte des 248 *vākya* de Vararuci n'a donc pas d'autre sens que des nombres, pas d'autre visée que le calcul, pas d'autre intérêt qu'astronomique. Il n'en reste pas moins que ces courtes chaînes de syllabes constituent des mots et, à part quelques cas, des fragments de phrases. Chacun de ces *vākyas* a donc un sens, mais qui pose une expectative, qui invite à chercher un complément pour former une phrase entière. Il est dès lors tentant d'exploiter cette expectative et d'insérer ce groupe de mots dans un ensemble signifiant plus long. Cet exercice s'apparente à un exercice apprécié des lettrés, appelé *samasyāpūraṇa*, la composition d'une strophe sur la base d'une proposition donnée. Un autre exercice apparenté est la composition d'une strophe de quatre vers en incorporant une citation d'un ouvrage bien connu. Un exemple est le *Candradūta* qui incorpore les derniers vers des strophes du *Meghadūta*. Le *Haricarita* accomplit ce tour de force en prenant chaque *vākya* de Vararuci comme un début de vers dans une composition dévotionnelle à l'adresse du dieu Kṛṣṇa.

Il nous est connu par un seul manuscrit qui dans son colophon ne donne pas de nom d'auteur. Son éditeur, C. Kunhan Raja, à titre d'hypothèse, a proposé d'attribuer cet ouvrage à un lettré du Kerala, Parameśvara Bhaṭṭa, de la famille des Payyur Bhaṭṭa. Son hypothèse, tout à fait vraisemblable, repose sur une strophe finale dans deux manuscrits d'un commentaire intitulé

śodhya 1600984 nombre de jours pour Kali 4383 (1282 A.D.). Kunhan Raja a pris 1741650 ce qui est valable pour quelque 350 années plus tard, bien après la date présumée du *Haricarita* qui doit être très proche de celle du *Vākyakaraṇa*. Nous avons donc préféré nous référer à ce dernier comme référence vraisemblable pour le *Haricarita*.

[5]Voir Alexander Jones, « The Development and Transmission of 248-Day Schemes for Lunar Motion in Ancient Astronomy » in *Archive for History of Exact Sciences*, vol. 29, n° 1, Berlin, 1983, pp. 1-36 ; K. Chandra Hari, « Vākyakaraṇa », in *Indian Journal of History of Science*, 36. 3-4, Delhi, 2001, pp. 127-149.

Sumanoramaṇī sur le *Meghadūta* :

anudinam abhinavarūpā sumanoramaṇīva jagati jayatitarām I
haricaritakāvyasahabhūr vyākhyāsau meghadūtasya II

« Telle une jeune femme qui plait par ses fleurs, à la beauté nouvelle chaque jour, cette glose [Sumanoramaṇī (qui plait aux lettrés)] du Meghadūta (Le nuage messager), sœur du poème Haricarita (La geste de Hari) a le plus grand triomphe dans l'univers. »

Dans un autre manuscrit de la même *Sumanoramaṇī*, qui n'a pas cette strophe, on en trouve un autre qui révèle le nom de l'auteur :

kaumārilācāryavarasya nāmnā vṛttena mūrtyāpi ca śaṅkarasya I
śiṣyeṇa sṛṣṭā parameśvareṇa vyākhyēyam enāṁ vimṛśantu santaḥ II

« Par Parameśvara disciple de Śaṅkara, le meilleur des maîtres dans la discipline de Kumārila, de nom, de conduite et même de corps, fut créée cette glose. Que les gens de bien réfléchissent sur elle. »

C. Kunhan Raja montre que ce personnage est, dans la famille des Payyūr Bhaṭṭa, Parameśvara I, fils de Ṛṣi et Gaurī, neveu de Bhavadāsa, élève de Śaṅkara, auteur d'ouvrages de *mīmāṁsā*, *Svaditaṅkaraṇī*, commentaire sur la *Nyāyakaṇikā*, un commentaire sur un *Nyāyasamuccaya* et, d'après le vers cité ci-dessus, auteur d'un *kāvya* intitulé *Haricarita*. Mais ce *Haricarita* est-il le texte que nous étudions ou un autre du même nom ? On souhaiterait un critère additionnel permettant de considérer qu'il s'agit bien du même texte. Un doute subsiste donc. Mais la présomption d'identification est renforcée par le fait que dans la même famille ont été composés deux ouvrages similaires en ce qu'ils sont faits de strophes sur la base des *vākya* de Vararuci, à savoir un ouvrage de *mīmāṁsā* intitulé *Kaumārilayuktimālā* et une *Vākyāvalī* qui, elle aussi, raconte l'histoire de Kṛṣṇa.

L'on n'a pas d'indication directe sur la date du présent *Haricarita*. Mais si son auteur est bien de la famille des Payyūr Bhaṭṭa qui s'est illustrée à la cour du Zamorin de Calicut au milieu du XIVe siècle, il doit se situer au début du XIVe siècle, peut-être à la fin du XIIIe, contemporain ou légèrement postérieur au *Vākyakaraṇa*.

Le premier trait qui caractérise ce *Haricarita* est l'habileté de l'insertion de la proposition astronomique. Dans sa nouvelle composition, le poète tantôt garde, tantôt change le sens du groupe de mots du *vākya*.

Un exemple facile où il le garde, est l'invocation à Sarasvatī qui ouvre le texte :

gīr naḥ śreyaḥ *paraṁ dadyād ajā sā caturākṛtiḥ I*
anurūpam ajaṁ vavre yā patiṁ caturānanam II 1 II

« Que **la Parole nous** gratifie du **bien suprême**, cette [déesse] non-née, à quatre visages, qui [se] choisit un époux digne d'elle, le dieu non-né, à quatre visages. »

Dans un autre cas le *vākya* est un groupe de mots avec un sens possible, mais qui est modifié par un nouveau découpage de la chaîne de syllabes. Par exemple *gṛhyāḥ surā rājñā* « les dieux doivent être adoptés par le roi » :

gṛhyāsur ārāj jñā*nasya phalaṁ tu svaḥstriyo 'ñjasā* |
janitvā bhuvi līlābhiḥ svābhir ārādhya keśavam || 7 ||

« **Puissent** les filles du ciel **recevoir vite** le fruit de **la connaissance**, en allant bientôt naître sur terre et après avoir propitié Keśava par leurs jeux. »

On remarque ici que le poète ne tient pas compte du *ḥ* de *gṛhyāḥ*, ce qui lui permet de faire un découpage tout à fait différent de la séquence initiale.

Dans le cas du *vākya* 154 *jitvāsya rathaḥ* « le char de celui-ci ayant vaincu » cité ci-dessus pour sa valeur numérique, une syllabe a été changée, conservant la même valeur numérique, *s* au lieu de *th* = 7, ce qui entraîne un sens nouveau, à savoir une évocation de l'attente de l'incarnation de Viṣṇu en Kṛṣṇa :

jitvāsya rasa*nādyakṣāṇy aniśaṁ vapuṣo budhāḥ* |
sphuranti sakṛd apy asminn āparokṣyaṁ bhaved iti ||

« Les sages qui pour toujours **ont vaincu, de ce** corps, **la langue** et autres organes des sens, exultent [à la pensée] qu'il puisse y avoir expérience directe de ce [Seigneur suprême], même une seule fois. »

Le poème suit le déroulement des exploits de Kṛṣṇa tels qu'ils sont donnés dans le *Bhāgavatapurāṇa*. Le poète répond donc à deux contraintes, la donnée des *vākyas* astronomiques et la donnée de l'autorité du *Purāṇa*. Il répond encore à une troisième inspiration, sa *bhakti* ou dévotion personnelle à l'égard de Kṛṣṇa. Et cela est le facteur qui apporte à l'œuvre toute son originalité et sa valeur poétique. Un seul exemple suffit à montrer le pas « littéraire » franchi entre le *vākya* 121 *pāpo'yaṁ niśi* « ce vilain la nuit ... » et son amplification en une strophe *śārdūlavikrīḍita* qui évoque gracieusement les bergères accourant au son de la flûte de Kṛṣṇa :

pāpo'yaṁ niśi *nityaśo nijagṛhe śete kvacid vañcayan*
yo māṁ mārahatām asau vidhivaśād āste vane'syāṁ niśi |
yāmy adyeti muhuḥ praṇamya śirasā viṣṇuṁ drutāyāṁ harer
gītau tatpratiśabdavad yuvatayo yānti sma tasyāntikam || 125 ||

« "D'ordinaire, **ce vilain, la nuit**, me trompant, moi qu'il a frappée d'amour, dort quelque part dans sa demeure ; cette nuit, par bonheur, il est resté dans le bois ; je vais y aller aujourd'hui", ce disant les jeunes filles qui s'étaient maintes fois prosternées, la tête au sol, devant Viṣṇu, au chant de rythme vif de Hari, telles des échos, allaient auprès de lui. »

La mention de l'adoration de Viṣṇu rappelle que le poème de *bhakti* oscille entre l'évocation du ciel et celle de la terre. Cette strophe juxtapose l'aventure terrestre des amoureuses et le culte du dieu. La strophe suivante, *vākya* 122 *dhanyo mānyo'ṁśe* « riche, respectable en partie ... », avec changement d'une voyelle et du sens, donne une explication philosophique de cet échange du divin et du terrestre : si le dieu unique multiplie son image pour chaque créature de ce monde, c'est que chacune n'est qu'un *śeṣa*, un reste surimposé sur sa nature de *śeṣin*, le possesseur et maître de tout ce qui ne peut être que reste englobé dans son essence infinie.

***dhanyo mānye śe**ṣabhāvam agamad gopikājane |*
pratyekaparikḷptābhir acyutaḥ svīyamūrtibhiḥ || 126 ||
« **Riche** [de tout, le *Śeṣin*], l'Impérissable prit l'état de **reste** dans **l'estimable** peuple des bergères, par les corps de lui-même qu'il créait pour chacune. »

Ce poème réunit de multiples savoirs, composition littéraire, astronomie, mythologie, philosophie de la dévotion. Y a-t-il un lien entre tous ces savoirs ? N'y a-t-il pas quelque disparate entre une table de positions lunaires et l'histoire de Kṛṣṇa ? Un lien très fort, capable d'inspirer une consécration dans un texte à double sens, nous paraît être tout simplement le mode de vie du brâhmane lettré et la très forte orthopraxie religieuse des Nambudiri du Kerala. Le brâhmane pieux a le devoir et la coutume de penser aux dieux en s'éveillant, de déterminer *tithi* et *nakṣatra* du jour, afin d'exécuter les rites et fêtes appropriés. Dévotion (*bhakti*) et rites, c'est-à-dire application du savoir rituel incluant la situation calendérique de chaque acte, sont son lot quotidien. Une mémoire exercée et une pratique constante lui ont donné un *saṁskāra* dont ce *Haricarita* est en quelque sorte une émanation. Dans la mémoire du brâhmane il est un bréviaire et un calendrier qui lui donne le moyen de faire un premier acte de piété et de déterminer la date et la liturgie du jour.

C'est, en pleine époque médiévale, une merveilleuse manifestation d'une culture scientifique, littéraire et religieuse d'un grand raffinement et d'une grande élévation.

Yantrarāja for Dāmodara
The Earliest Extant Sanskrit Astrolabe

Sreeramula Rajeswara Sarma
Formerly Professor of Sanskrit
Aligarh Muslim University

1.0 Introduction

The astrolabe enjoyed great esteem, as no other instrument did, in the medieval world between Geoffrey Chaucer's England and Firūz Shāh Tughluq's Delhi. Its origins are, however, rather obscure. All that is known with certainty is that it was designed sometime after 150 BC when the stereographic projection, on which its construction is based, was said to be have been invented by Hipparchus and that almost all its main components were fully developed by 530 AD when the earliest available work on it was composed by John Philoponus in Greek.[1] The Islamic world preserved the Greek science of the astrolabe, as it did the other areas of Greek learning, elaborated upon it and disseminated it westwards and eastwards.

Again, we do not have any definite information when the astrolabe was introduced into India. Al-Bīrūnī, who wrote extensively on the astrolabe, may have introduced it to his Hindu interlocutors in north-western India in the first quarter of the eleventh century. There are stray references to Muslim scholars, who migrated from Central Asia in the subsequent centuries, employing the astrolabe for astronomical and astrological purposes in Delhi.

Definite evidence is available from the fourteenth century onwards. Contemporary chronicles report that Sulṭān Fīrūz Shāh Tughluq (r.1351-1388) got several astrolabes manufactured at Delhi and sponsored the composition

[1]For an English translation of this work, see R. T. Gunther, *The Astrolabes of the World*, Oxford 1932, reprint: London 1976, vol. I, pp. 61-81.

of manuals on the astrolabe both in Persian and in Sanskrit.[2] In the following centuries, especially in the seventeenth, numerous astrolabes were produced in India with Arabic/Persian legends. These are generally classified as Indo-Persian astrolabes.[3]

Hindu and Jain astronomers were also highly enthusiastic about this exotic, but versatile, instrument which they hailed in Sanskrit as *Yantrarāja*, 'king of instruments'. They 'appropriated' it by composing more than a dozen Sanskrit manuals on it between the fourteenth and eighteenth centuries.[4]

The very first Sanskrit manual on the astrolabe was composed in 1370 by a Jain monk called Mahendra Sūri, who was a leading astronomer at the court of Firūz Shāh at Delhi. This work, entitled *Yantrarāja*,[5] consists of five chapters. The first chapter provides various trigonometric parameters and tables needed for the construction of the astrolabe. The second chapter discusses the components of the astrolabe. The construction of the common northern astrolabe (*saumya-yantra*) and other variants is described in the third chapter, while the next one deals with the method of verifying whether an astrolabe is properly constructed or not. The final chapter discusses the use of the astrolabe as an observational and computational device and dwells on problems in astronomy and spherical trigonometry that can be solved by means of the astrolabe. Mahendra Sūri's pupil Malayendu Sūri wrote a detailed commentary on the *Yantrarāja*. There survive today at least one hundred manuscript copies of the *Yantrarāja*, attesting to its great popularity.

Persian chronicles narrate that Firūz Shāh was interested not only in the commonly used northern astrolabe, but also in its variants, viz., the southern astrolabe and the composite astrolabe which is a combination of the two, and that he commissioned a very splendid specimen of the composite as-

---

[2]See Sreeramula Rajeswara Sarma, 'Sulṭān, Sūri and the Astrolabe' in: idem, *The Archaic and the Exotic: Studies in the History of Indian Astronomical Instruments*, New Delhi 2008, pp. 179-198.

[3]On Indo-Persian astrolabes, see Sreeramula Rajeswara Sarma, 'Astronomical Instruments in Mughal Miniatures' in: idem, *The Archaic and the Exotic*, op. cit., pp. 76-121; 'The Lahore Family of Astrolabists and their Ouvrage,' ibid., pp. 191-222; 'The *Ṣafīḥa Zarqāliyya* in India,' ibid., 223-239; *Astronomical Instruments in the Salar Jung Museum*, Hyderabad 1996; *Astronomical Instruments in the Rampur Raza Library*, Rampur 2003.

[4]These texts are discussed in Sreeramula Rajeswara Sarma, 'Yantrarāja: the Astrolabe in Sanskrit' in: idem, *The Archaic and the Exotic*, op. cit., pp. 240-256.

[5]Mahendra Sūri, *Yantrarāja of Mahendra Sūri together with the commentary of Malayendu Sūri, and Yantraśiromaṇi of Viśrāma*, ed. Kṛṣṇaśaṃkara Keśavarāma Raikva, Bombay 1936.

trolabe which was named Fīrūz Shāh's Astrolabe (*asṭurlāb-i Fīrūz Shāhī*). The difference between these varieties is as follows. In the northern astrolabe (Arabic: *asṭurlāb shumālī*, Sanskrit: *saumya-yantra*), the rete contains pointers for stars situated between the north celestial pole and the Tropic of Capricorn. As against this, the rete of the southern astrolabe (Arabic: *asṭurlāb janūbi*, Sanskrit: *yāmya-yantra*) has pointers for stars that lie between the south celestial pole and the Tropic of Cancer. The third variant combines the features of these two and is called the north-south astrolabe (Arabic: *asṭurlāb shumālī wa janubī*, Sanskrit: *miśra-yantra*). The two variants are actually theoretical constructs and have no real practical value. This is evident from the fact that out of some three thousand and odd Islamic, European and Sanskrit astrolabes that are extant today, the southern and north-south variants account for less than a score of specimens. Even so, most of the Sanskrit writers discussed all the three varieties. In fact, in 1423 Padmanābha composed an exclusive manual on the southern astrolabe under the title *Yantrarājādhikāra*.[6]

The next writer of note is Rāmacandra Vājapeyin[7] who devoted the major part of his *Yantraprakāśa* (1428)[8] to the astrolabe and declared that, if one knew the science of the astrolabe well, the entire universe would become comprehensible like the myrobalan fruit on one's own palm (1.9: *yasmin karāmalakavad vidite viditaṃ bhaved viśvam*).

The astrolabe received a great impetus from Sawai Jai Singh in the early eighteenth century. Although he preferred huge masonry instruments for astronomical observations, he had great esteem for the astrolabe on which he composed a manual in Sanskrit prose under the title *Yantrarājaracanā*.[9] Even after Jai Singh's time, the astrolabe continued to be discussed in several Sanskrit works.

These Sanskrit manuals on the astrolabe must naturally have been accompanied by the production of Sanskrit astrolabes, i.e., astrolabes with legends and numbers engraved in Sanskrit language and Devanāgarī script. In the

---

[6]In his 'Early History of the Astrolabe in India,' *Indian Journal of History of Science*, 32 (1997) 199-295, Yukio Ôhashi offers an excellent edition, translation and mathematical commentary of Padmanābha's *Yantrarājādhikāra*.

[7]Sreeramula Rajeswara Sarma, 'On the Life and Works of Rāmacandra Vājapeyin' in: *Śrutimahatī: Glory of Sanskrit Tradition* (Professor R. K. Sharma Felicitation Volume), Delhi 2008, vol. 2, pp. 645-661.

[8]Together with an auto-commentary, available in MS G-1363 of the Asiatic Society, Kolkata, and MS 975/1886-92 of the Bhandarkar Oriental Research Institute, Pune.

[9]Kedāranātha Jyotirvid (ed), *Yantrarājaracanā of Jayasiṃhadeva & Yantraprabhā of Śrīnātha*, Jaipur 1953.

course of my project of surveying the extant specimens of pre-modern Indian astronomical and time-measuring instruments,[10] I have located nearly one hundred Sanskrit astrolabes in various museums and private collections in India, Europe and the US.

There is a basic difference between these Sanskrit astrolabes and the Islamic astrolabes. In Islamic culture, astrolabe making was not just a craft but a learned profession; the astrolabists were not mere metal workers, but also scholars well read in the literature on the instruments, well versed in astronomy, spherical trigonometry and other sciences. Here the same person prepared the technical design and then executed it from brass sheets. In the Hindu context, the technical design was drawn by the upper caste astronomer, and the actual manufacture was done by the lowly brass worker who may not even be literate. He managed to draw the lines and curves reasonably well, but often made mistakes in orthography.

Nevertheless, the large number of extant Sanskrit texts and Sanskrit astrolabes show the importance given to the astrolabe by the Hindu astronomers. We have noted that production of Sanskrit manuals commenced in the fourteenth century. Sanskrit astrolabes also must have been produced in that century, but those that survive are only from the seventeenth century onwards. I have identified the following ten astrolabes from the seventeenth century. Most of these were produced in Gujarat.

1. 1605 Made for Dāmodara, private collection, Brussels.
2. 1618 Anon, Raja Dinkar Kelkar Museum, Pune.
3. 1625 Made by Cakrapāṇi[11] (present location unknown).
4. 1638 Made by Murāraji for Haranatha, Bhandarkar Oriental Research Institute, Pune.
5. 1642 Made by Kalyāṇa of Girinārāyaṇa-jñāti for Puruṣottama (present location unknown).
6. 1643 Made for Maṇirāma, Royal Scottish Museum, Edinburgh.[12]

[10]On this project, see Sreeramula Rajeswara Sarma, 'Indian Astronomical and Time-Measuring Instruments: A Catalogue in Preparation' in: idem, *The Archaic and Exotic*, op. cit., pp. 19-46.

[11]It was auctioned in 2002; cf. Skinner, Bolton, USA, Auction Catalogue 'Science & Technology featuring Mechanical Music,' April 13, 2002, no. 244, pp. 38-39; only the mater is extant.

[12]Cf. Sreeramula Rajeswara Sarma, 'Yantrarāja at Edinburgh: On a Sanskrit Astrolabe made for Maṇirāma in AD 1644,' to appear in S. R. Sarma & Gyula Wojtilla (ed), *Scientific Literature in Sanskrit*, (Proceedings of the 13th World Sanskrit Conference, Section 8), Motilal Banarsidass, Delhi, p. 77-110.

7. 1651 Made for Jīvatāpana, Oriental Institute, Vadodara.[13]
8. 1669 Made for Rāghavajit (present location unknown).[14]
9. 1673 made for Indrajī, Pitt Rivers Museum, Oxford.[15]
10. Anonymous, not dated, but attributable to the seventeenth century, Sanskrit University, Varanasi.[16]

2.0 Dāmodara's Astrolabe

The earliest of these, in fact, the earliest extant Sanskrit astrolabe, is the subject of this paper. It was produced at Ahmedabad for Dāmodara, son of Caṇḍīdāsa, and is now in a private collection in Brussels. There has been some controversy about its date and authenticity. I shall first describe the astrolabe as it is available today, comparing its construction, where necessary, with the other seventeenth century Sanskrit astrolabes and also with Indo-Persian astrolabes. I shall then explain the controversy, and finally establish its authenticity.

Since this is the earliest extant Sanskrit astrolabe, some extracts from the earliest Sanskrit manual on the astrolabe, viz. the *Yantrarāja* of Mahendra Sūri, dealing with the construction of the astrolabe, are given in the appendix. As stated above, this text was immensely popular in India. It will be shown that its influence is clearly discernable in the Sanskrit astrolabes of the seventeenth century, including the present astrolabe.

2.1 The Front of the Astrolabe

The present astrolabe is a large massive piece with a diameter of 276 mm, a height of 380 mm and a thickness of 10.3 mm. The body is surmounted by an elaborately pierced suspension bracket or crown (*kirīṭa*) with a very

[13]Cf. Sreeramula Rajeswara Sarma, *Sanskrit Astronomical Instruments in the Maharaja Sayajirao University of Baroda*, M.S. University Oriental Series No. 24, Oriental Institute, Vadodara, 2009, pp. 23-32, Figs. 11-19.

[14]In 1936, it was in the possession of Raikva, who published 13 detailed illustrations of this astrolabe in his edition of Mahendra Sūri's *Yantrarāja* (see n. 5 above) immediately after his introduction. Its present location is unknown.

[15]Cf. R. T. Gunther, *Early Science in Oxford*, vol. II: Astronomy, Oxford 1923, pp. 187-199, esp. 198; idem, *The Astrolabes of the World*, op. cit., p. 211, Fig. 110.

[16]Cf. Sreeramula Rajeswara Sarma, 'Kaṭapayādi Notation on a Sanskrit Astrolabe' in: idem, *The Archaic and Exotic: Studies in the History of Indian Astronomical Instruments*, New Delhi 2008, pp. 257-272.

wide base that stretches to about 106 degrees of arc (see Figure 1[17]). The crown is worked *à jour* to produce an ornate pattern with a human figure entwined in vines and leaves. The surface of the crown is, however, plain and undecorated, and no attempt was made to smoothen and polish the sharp edges. There is a hole at the top of the crown through which passes a shackle (*kaṇṭaka*) with ornate terminals. A small ring (*mudrikā*) passes through the shackle. The shackle and the ring make it possible to suspend the astrolabe vertically and to rotate it all around. The ring is, however, much too small for this large astrolabe. Both the shackle and the ring have diamond-shaped cross-sections as in the contemporary Indo-Persian astrolabes. At the top of the crown, immediately below the hole is an inscription in Sanskrit in Devanāgarī letters which reads *śrī-divyacakṣuṣe namaḥ*, 'salutation to the divine eye,' i.e. the sun.

The main body of the astrolabe, consisting of a thick brass disc with an upraised rim and the crown, was cast in one piece. It is called mater (*koṣṭhakāgāra*, lit. store-house, repository) because the recess inside the upraised rim accommodates several latitude plates and above them the perforated star map called rete. The upraised rim carries the degree scale in two columns or bands. The narrow inner band is graduated in single degrees of arc; the wider outer column is divided into groups of 3 degrees each and labelled with very elegant Devanāgarī numerals of western Indian style.

2.2 The Back of the Astrolabe

The back of the astrolabe (*yantra-pṛṣṭha*) (Figure 2) carries the diopter or alidade with which the celestial bodies are viewed and the angle of their elevation or altitude is measured. The alidade, which is pivoted to the centre, is a straight metal bar without any counter-change and is 265 mm long. It has pointed ends. A sighting tube is attached to it by means of two upright supports. The tube measures 290 mm and is slightly longer than the diameter of the astrolabe. This tube facilitates the sighting of the heavenly bodies and is a typical feature of Sanskrit astrolabes.

Islamic astrolabes, whether in India or outside, do not have sighting tubes attached to their alidades. The same is the case with European astrolabes.

[17]Photo credits: Figure 2 is from Christie's, South Kensington, *Time Measuring Instruments from THE TIME MUSEUM*, for sale by Auction, Thursday 14 April 1988, Catalogue, no. 157, pp. 98. Figures 3 and 4 are by the courtesy of Anthony Turner. The remaining photos are by me.

The Sanskrit astrolabes, however, carry a sighting tube in most of the cases. In India, the sighting tube was used as an independent device to view celestial bodies.[18] It was called *Nalaka-* or *Nālikā-yantra* and was described in several Sanskrit texts.[19] Apparently Hindu astronomers decided, at quite an early period, to combine the already known sighting tube with the newly imported astrolabe. This, no doubt, facilitates the sighting, but prevents the calibration of the upper surface of the alidade and thus divests it off the trigonometric functions envisaged for the alidade in Islamic tradition. No Sanskrit text appears to mention this innovation.

The surface of the back is divided into four quadrants by the diameters drawn in the south-north and east-west directions. The edge of the entire surface is occupied by a degree scale. As in the front, here also the inner band of the scale is graduated in single degrees and the outer band is divided into groups of three degrees. But here the groups of three degrees are labelled separately for each quadrant, starting from the east or west points and proceeding towards the south or west points. The degrees scales in the upper half, starting from the east and west points and reaching the south point, are useful for measuring the angle of altitude of a heavenly body situated to the east or west of the meridian. Those in the lower half can also be used for the same purpose, but are actually redundant. Islamic astrolabes carry cotangent scales in the lower half, but these are not met with in Sanskrit astrolabes.

The spaces inside the circular scale in the four quadrants are filled with different trigonometric graphs. The upper left quadrant contains a sexagesimal sine graph, with sixty horizontal parallel lines drawn at equal intervals. These are numbered serially, three at a time, on the right along the vertical radius, starting from the centre and proceeding upwards as 9, 12, 15, ... 54, 57, 60. While the angle of altitude is measured on the circular degree scale on the left edge, the sine of the angle can be directly read off the vertical scale on the right. An arc is drawn at 24° inside the quadrant to mark the obliquity of the ecliptic. The quadrant is divided by three radii into three sectors of 30° each. In these sectors are engraved the numbers 1 to 12 in the following manner. In the first row immediately next to the arc of the obliquity are the numerals 1, 2, 3, written in each sector. The tops of these numerals are towards

[18] It was also known to the Greeks and Arabs; cf. Peter Schmalzl, *Zur Geschichte des Quadranten bei den Arabern*, München 1929; reprint in: *Islamic Mathematics and Astronomy*, vol. 90 (Astronomical Instruments and Observatories in the Islamic World: Text and Studies VI), Frankfurt 1998, pp. 189-331, esp. 218-219.

[19] Cf. R. N. Rai, 'Astronomical Instruments,' *Indian Journal of History of Science*, 20 (1985) 308-336, esp. 333-336: The Nalaka-yantra.

Figure 1: The Front of the Astrolabe

Figure 2: The Back of the Astrolabe

the degree scale in the circular edge. In the next row are the numerals 6, 5, 4 (as read from left to right). These are upside down (i.e., the tops are towards the centre). In the third row are 7, 8, 9 with orientation as in the first row. The fourth row has numbers 12, 11, 10, with orientation as in the second row.[20]

The upper right quadrant contains a series of curves for unequal hours, which are numbered from 1 to 12. Of the seventeenth-century Sanskrit astrolabes, only the astrolabe made for Indrajī in 1673 has this feature. Otherwise it occurs rarely in Sanskrit astrolabes and also in Indo-Persian astrolabes.[21]

In the lower half are engraved shadow squares, on the left for a gnomon of 12 digits, and on the right for a gnomon of 7 digits. The scales are numbered, but there are no labels.

The seventeenth century Indo-Persian astrolabes contain, besides these trigonometric graphs, much astrological data on the back. The Sanskrit astrolabes rarely follow this practice, and the present astrolabe is no exception.

On the inner side of the mater, on the meridian line below the centre and somewhat close to the rim, is a pin affixed at right angles to the surface of the mater. This pin holds the latitude plates firmly in position and prevents them from rotating around the centre.

On the same inner side, in the upper half is an inscription, mentioning the name of the patron who caused this astrolabe to be made and the date of manufacture. The inscription will be discussed below in section 3.

2.3 The Latitude Plates

The recess formed by the upraised rim of the mater accommodates six thin circular plates. These are called in Sanskrit *akṣa-patra* or *akṣāṃśa-patra*, because each face of the plate carries stereographic projections of the heavens

[20]Probably these numbers refer to the 12 signs of the zodiac. In a sine quadrant made by Jamāl al-Dīn Muḥammad 'Alī al-Ḥusaynī in AH 1273 (AD 1856-57), in the space between the arc of the quadrant and the corner of the plate are engraved the names of twelve signs in the same order. However, I am unable to explain the arrangement. Does this have to do with some type of classification of the signs? The quadrant is now with the Rampur Raza Library; cf. Sreeramula Rajeswara Sarma, *Astronomical Instruments in the Rampur Raza Library*, op. cit., pp. 76-77 and Fig. 9.1 on p. 75.

[21]Mahendra Sūri, *Yantrarāja* 3.3 (see Appendix) mentions that the quadrants on the back should be filled with a sine graph, a declination graph and shadow squares. So does Rāmacandra Vājapeyin, *Yantraprakāśa* 2.7d: *kecit procur apakramāṃśa-guṇa-śaṅkvābhā-sthitiṃ tatra ca*, 'Some prescribe there the setting up (*sthiti*) of the declination (*apakrama*) [graph], sine (*guṇa*) [graph] and gnomon shadow (*śaṅkvābhā*) [squares].'

related to a specific terrestrial latitude (*akṣa, akṣāṃśa*).[22]

These six plates (Figures 3 and 4) have a diameter of 252 mm. There is a hole at the centre through which passes the pin that holds together all the components of the astrolabe. This pin, like the centres on all the plates, defines the north celestial pole. Below the centre, on the lower half of the meridian, there is another hole. When the plates are stacked together inside the recess, the short pin projecting from the lower half of the mater passes through this second hole in the plates and holds them firmly in position.

On both faces of all the six plates, the projections are drawn in the following manner. First the four cardinal directions are marked on the face of the plate, by means of the east-west diameter (*pūrvāparasūtra*) and south-north diameter; the latter represents the meridian line (*khamadhya-sūtra*) or the 12 o'clock line (*madhyāhna-rekhā*).

Then taking the radius of the plate as 30 units, three concentric circles are drawn with radii of 30, 19;38 and 12;51 units respectively. The outermost circle with the radius of 30 units forms the periphery of the plate and defines the tropic of Capricorn (*makarāhorātra-vṛtta*). The middle circle with the radius of 19;38 units represents the celestial equator. The inner circle with the radius of 12;51 units is the tropic of Cancer (*karkāhorātra-vṛtta*).

Against this grid is drawn the local or oblique horizon (*kṣitija-vṛtta*). Above the arc of the horizon circle, the equal altitude circles or almucantars (*unnata-vṛtta*) are drawn for each 3 degrees of arc,[23] with varying radii and varying centres (which are all situated on the meridian).[24] All the equal altitude circles are numbered on the right as well as on the left, as 3, 6, 9 ...90. The rows of the numbers form a pattern that looks like a rounded Roman character 'M', as in the Indo-Persian astrolabes.

It is worth noting that no azimuth lines (*digaṃśa-vṛttas*) are drawn on any

[22]On the construction of the latitude plates, see Mahendra Sūri, *Yantrarāja* 3.5-7, Appendix.

[23]In Islamic tradition, astrolabes are classified according to the number of almucantar circles drawn on the latitude plates. Mahendra Sūri also mentions this classification at *Yantrarāja* 3.1, (see Appendix). Therefore this astrolabe belongs to the tripartite (Arabic: *thulthī*) class. The tripartite arrangement can be seen also in the degree scale on the front and back of the mater, in the vertical scale of sines in the upper left quadrant on the back, and in the subdivision of the signs in the ecliptic ring on the rete.

[24]Mahendra Sūri teaches the method of computing the radii (*vyāsārdha*) and the eccentricities (*kendra*) of different altitude circles in his *Yantrarāja* 1.18-21. In his commentary (pp. 19-25) Malayendu Sūri provides the values of the radii and eccentricities, at six degree intervals, for six towns, viz. Tilaṅga (lat. 18°), Tryambaka (21°), Aṇahillapattana (lat. 24;30°), Tīrabhukta (lat. 27°), Dillī (28;39°) and Nepālapura (31°).

plate, which are usually to be met with in Islamic astrolabes. In contrast, Sanskrit astrolabes rarely carry azimuth lines.

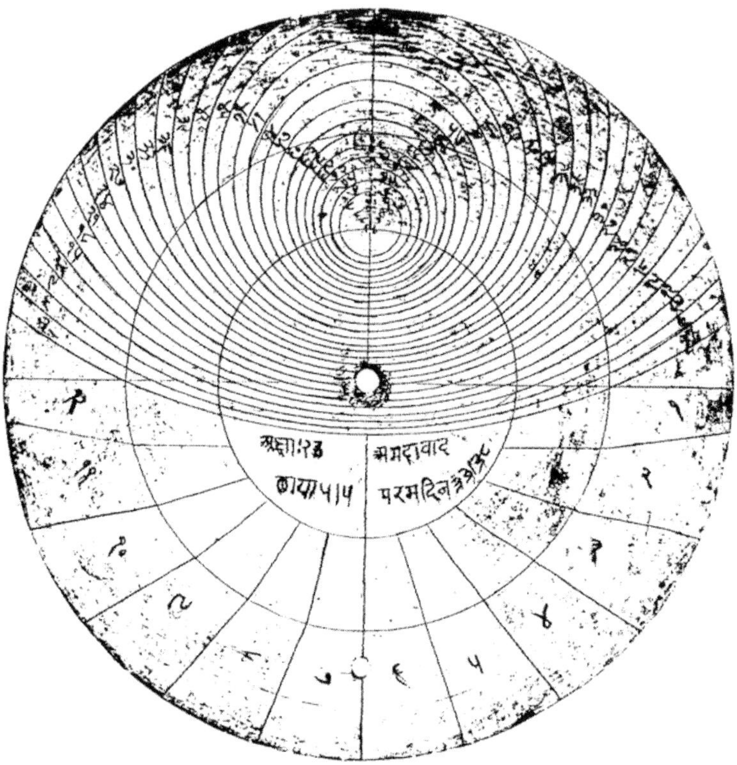

Figure 3: Plate for the Latitude of Ahmedabad

In the lower half of the plate, the arcs of the three circles representing the tropic of Capricorn, the equator and the tropic of Cancer lying below the oblique horizon are divided into 12 parts each. The corresponding points of division on the three arcs are joined, not by arcs as is the general practice, but by two straight lines. That is to say, the first point of division on the tropic of Cancer is joined to the first point on the equator by a straight line, and the first point on the equator is joined to the corresponding point of the tropic of

Capricorn by another line. These lines are designated as unequal hour lines or seasonal hour lines.[25] The unequal hours (*viṣama-horā*) are numbered with the numerals 1 to 12, starting at the western horizon and going up to the eastern horizon. The Hindus and Jains in India did not reckon time in unequal or equal hours, but in equal *ghaṭīs* (of 24 minutes each) and began the day with the sunrise. Lines for equal *ghaṭīs*, as commensurate with the actual practice in India, can easily be drawn, but strangely this has never been attempted in any Sanskrit astrolabe, although some Indo-Persian astrolabes have such lines.

Mahendra Sūri mentions only the curves for unequal hours as counted from the western horizon.[26] But he also teaches a simple method for measuring time in equal *ghaṭīs* as follows. First measure the sun's altitude with the alidade on the back of the astrolabe. Note the sun's longitude (*ravyaṃśaka*) for the day from some almanac and locate that point (S) on the ecliptic in the rete. Rotate the rete in such a way that the point S touches the eastern horizon. Note where the first point of Capricorn (*mṛgāsya*) touches the circular scale on the rim. Then rotate the rete once again so that S rests on the altitude circle corresponding to the sun's altitude jut measured. Note again where the first point of Capricorn touches the circular scale on the rim. The interval between the two positions is in degrees of arc. Divide it by 6. The result in *ghaṭīs* is the time elapsed since sunrise.[27]

---

[25]The unequal hours are obtained by dividing the duration of the daytime and that of the night separately by 12. Except on the days of equinox, such hours vary from the day to the night, and from one day to the next, according to the seasons.

[26]Mahendra Sūri, *Yantrarāja* 3.9 (see Appendix).

[27]Ibid., 5.3-4ab:

ravyaṃśakaṃ prāgapare ca bhūje dhṛtvonnatāṃśoparige kṛte'smin |
dvisthānasaṃlagnamṛgāsyamadhye kālāṃśakaiḥ paṅkti 10 palapramāṇaiḥ ||
ṣaḍbhir vibhaktair divasasya yātaṃ śeṣaṃ ca ghaṭyādi parisphuṭaṃ syāt |

"[Rotate the rete in such a way that] the sun's [longitude in] degrees (*ravy-aṃśaka*) [as seen in the ecliptic, on the desired day] falls on the eastern horizon (*bhūja*) [if it is the forenoon] or on the western horizon [if it is the afternoon, and then rotate again so that the longitude] falls on the altitude degrees. The time degrees between the two positions of the first point of Capricorn (*mṛgāsya*), each of which equals 10 *palas*, are divided by 6. The quotient in *ghaṭīs* etc. is the time elapsed or the time to come in the day [in the forenoon and in the afternoon respectively].

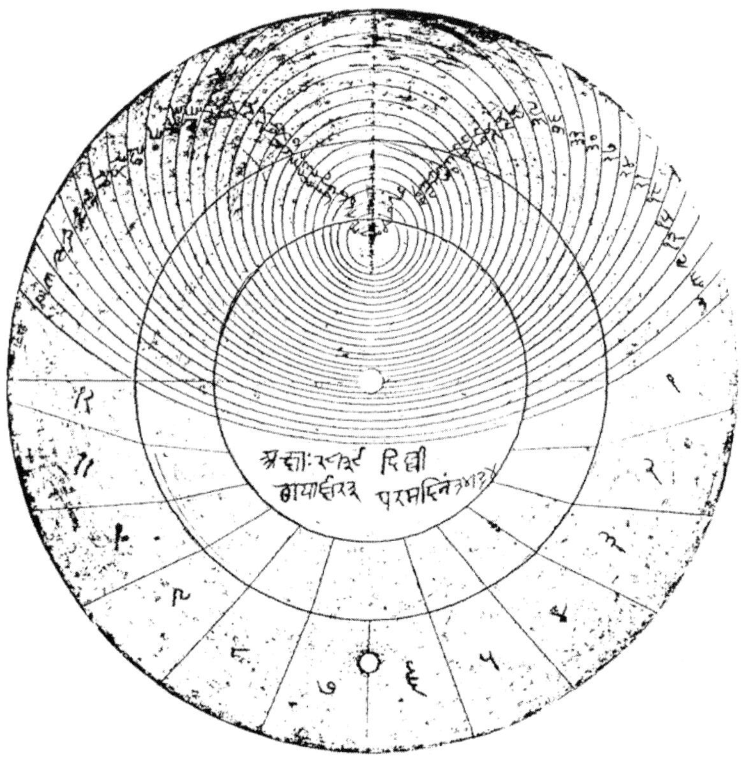

Figure 4: Plate for the Latitude of Delhi

Finally, in the space between the oblique horizon and the tropic of Cancer on each face are engraved four items of data regarding the projections drawn on this particular face, viz. (i) the name of a town, (ii) its latitude (*akṣāḥ*, φ) in degrees and minutes, (iii) the length of the midday equinoctial shadow (*chāyā*, tan φ), in *aṅgulas* and *vyaṅgulas*, thrown by a gnomon of 12 *aṅgulas* and (iv) the length of the longest day at this latitude (*paramadina*)[28]

[28]Christie's: South Kensington, *Time Measuring Instruments from THE TIME MUSEUM*, for sale by Auction, Thursday 14 April 1988, Catalogue, no. 157, pp. 98-99, mentions erroneously that the plates carry the length of **half** the longest day, but it is actually the length of the full day.

in *ghaṭīs* and *palas*. In fact, (iii) the length of the midday equinoctial shadow and (iv) the duration of the longest day are dependent on (ii) the latitude and are its functions. Sanskrit astronomical texts provide formulas for converting any one of the three values into another.[29] Islamic astrolabes generally mention only the latitude (*al-arḍ*) and the longest day (*al-sā'āt*). Mahendra Sūri indeed recommends that these four items be engraved on the astrolabe plates.[30]

The relevant inscriptions are reproduced below. The serial numbers for the plates and the designation of the two faces of the plates as 'a' and 'b' respectively have been added by us.

1a	*akṣāḥ*	*18*	*vījāpura*	*chāyā*	*3/45*	*paramadinaṃ*	*32/52*
1b	*akṣāḥ*	*22/30*	*ujjani*	*chāyā*	*4/9*	*paramadinaṃ*	*33/34*
2a	*akṣāḥ*	*20/30*	*bahrānapura*	*chāyā*	*4/30*	*paramadinaṃ*	*33/16*
2b	*akṣāḥ*	*23*	*amadāvāda*	*chāyā*	*5/5*	*paramadina*	*33/38*
3a	*akṣāḥ*	*25/52*	*kāśī*	*chāyā*	*5/45*	*dinamānaṃ*	*34/5*
3b	*akṣāḥ*	*27/40*	*ḍhākā*	*chāyā*	*6/20*	*paramadina*	*34/24*
4a	*akṣāḥ*	*25/56*	*yodhapura*	*chāyā*	*5/51*	*paramadinaṃ*	*34/8*
4b	*akṣāḥ*	*31/50*	*lāhora*	*chāyā*	*7/30*	*paramadina*	*35/20*
5a	*akṣāḥ*	*26/24*	*āgarā*	*chāyā*	*6/*	*paramadinaṃ*	*34/12*
5b	*akṣāḥ*	*28/39*	*dillī*	*chāyā*	*6/23*	*paramadinaṃ*	*34/34*
6a	*akṣāḥ*	*29/40*#	*mulatāṇa*	*chāyā*	*6/47*	*paramadi°*	*34/44*
6b	*akṣāḥ*	*35/20*	*kaśmīra*	*chāyā*	*8/6*	*paramadi*	(not filled)

It looks as if the latitude value of Multan was first engraved as 39/40 and then corrected to 29/40. The values given for the equinoctial shadow and the longest day match the latitude value of 29/40. Pingree read the latitude as 39/40 and this wrong reading was reproduced by Turner in his catalogue.

This material is arranged and interpreted in Table 1. The towns are re-arranged according to increasing latitude.

Interestingly these plates are town specific and not latitude or climate specific. Except in the case of Dhaka, the latitude values are reasonably close to modern values. The Islamic astrolabes generally contain, besides the plates calibrated to specific latitudes, one more plate, one face of which is designated as *ṣafīḥa mīzān al-'ankabūt* (plate for the measures on the rete) and

[29] Mahendra Sūri, *Yantrarāja* 3.25-27, gives formulas for converting local latitude into the duration of the longest day and the longest day into equinoctial shadow.

[30] Cf. Mahendra Sūri, *Yantrarāja* 3. 8 (see Appendix).

S. No.	Town Name as engraved	Modern Name	Latitude (*akṣāḥ*)	Modern Coordinates	Equinoctial shadow (*chāyā*)	Longest Day (*paramadina*) in *ghaṭīs* and *palas*	Longest Day in hours and minutes
1a	bījāpura	Bijapur	18	16;50° E 75;47° N	3;45	32;52	13;08.48
2a	bahrānapura	Burhanpur	20;30	21;17° N 76;16° E	4;30	33;16	13;18.24
1b	ujjani	Ujjain	22;30	23;09° N 75;43° E	4;09	33;34	13;25.36
2b	amadāvāda	Ahmedabad	23	23;03° N 72;40° E	5;05	33;38	13;27.12
3a	kāśī	Varanasi	25;52	25;20° N 83;00° E	5;45	34;05	13;38
4a	yodhapura	Jodhpur	25;56	26;18° N 73;04° E	5;51	34;08	13;44.12
5a	āgarā	Agra	26;24	27;10° N 78;05° E	6;00	34;12	13;40.48
3b	ḍhākā	Dhaka?	27;40	23;43° N 90;26° E	6;20	34;24	13;45.36
5b	dillī	Delhi	28;39	28;38° N 77;12° E	6;23	34;34	13;49.36
6a	mulatāna	Multan	29;40	30;12° N 71;31° E	6;47	34;44	13;53.36
4b	lahora	Lahore	31;50	31;37° N 74;26° E	7;30	35;20	14;08
6b	kaśmīra	Kashmir (actually) Srinagar	35;20	34;06° N 74;51° E	8;06	—	—

Table 1: Geographical Data engraved on the Plates

the other as *ṣafīḥa al-afāqiyah* (plate of horizons). The former is a projection for the latitude which is the complement of the obliquity of the ecliptic, i.e. 90-23;50 = 66;30° or roughly 66°. This plate enables us to measure the longitudes and latitudes of all the stars marked on the rete. The plate of horizons contains the projections of families of horizons at several latitudes and is used for determining the times of sunrise and sunset at latitudes other than one's own, or to determine the latitude from the time of sunrise or sunset. The present astrolabe does not have such a plate.

In sum, the plates in this Sanskrit astrolabe are generally calibrated as those in Islamic astrolabes, but differ from the latter in three details, viz. there is no plate for the *ṣafīḥa mīzān al-'ankabūt* and the *ṣafīḥa al-afāqiyah*, they do not have azimuth lines, and on each face is engraved the length of the equinoctial shadow in addition to the usual data.

On all the six plates, the almucantar circles are well drawn. However, the calligraphy of the legends and numerals is not the same as that on the mater and is considerably inferior. The workmanship of these plates is also rather inferior to that of the mater.

2.4 The Rete

On the top of the six latitude plates rests the rete (*bha-patra*, *bhacakra-patra*), an openwork disc, somewhat thicker than the latitude plates (Figure 5). It carries the stereographic projection of the sphere of the fixed stars.[31] With its 250 mm diameter, it is slightly smaller than the latitude plates. As on the latitude plates, here also the three concentric circles are drawn to represent the tropics and the celestial equator. The ecliptic (*krānti-vṛtta*) is drawn with radius of 21;26 units from a centre situated on the meridian at 8;34 units south of the centre of the plate,[32] so that it touches the tropic of Capricorn at the top and the tropic of Cancer below.

Some space is added inside the three circles of Capricorn, equator and ecliptic so that they have the shape of bands. The circular bands of Capricorn and equator are represented almost fully in the rete. The former is interrupted where the ecliptic touches it tangentially, while the band of the

[31] Mahendra Sūri explains the construction of the rete in his *Yantrarāja* 3.10-16, Appendix.

[32] Mahendra Sūri, *Yantrarāja* 1.66:

bhāgā bhacakrakendrasyāṣṭau liptā vedavahnayaḥ |
vyāsārdhasya dharādoṣo rasanetrāṇi ca kramāt ||

'The eccentricity (*bhacakra-kendra*) is 8 degrees and 34 minutes (i.e. actually 8;34 units where each unit is 1/30th of the astrolabe's radius) and the radius (*vyāsārdha*) 21;26 [units].'

equator has interruptions at two symmetrical places in the east and west. The ecliptic band is shown completely. The three bands are held together by the horizontal equinoctial bar which is without counter-change. The band of Capricorn and the lower segment of the equatorial band are joined together by a vertical bar on the meridian. This bar carries a short copper knob for rotating the rete around the centre.

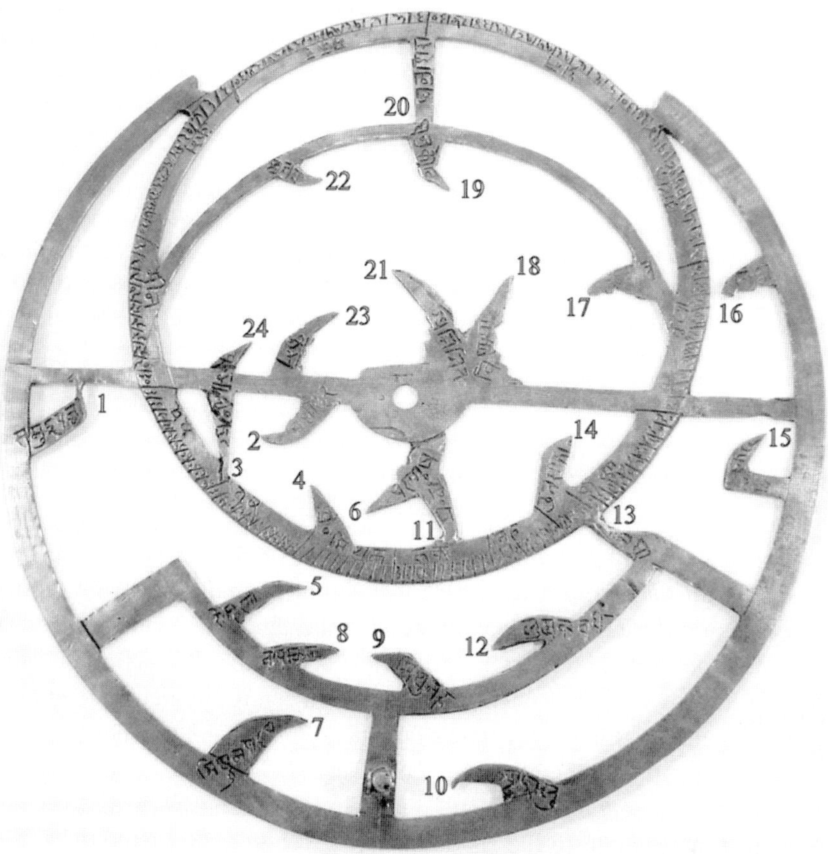

Figure 5: The Rete of the Astrolabe

The ecliptic circle is divided into the 12 signs of the zodiac, in unequal divisions in proportion to the rising times of the signs at the equator (*nirakṣodaya* or *laṅkodaya*), i.e. right ascensions. On each sign the respective Sanskrit names are engraved somewhat carelessly as *maṣa* (sic! read *meṣa*), *vaṣa* (*vṛṣa*), *mithana* (*mithuna*), *karka*, *siṃha*, *kanyā*, *tulā*, *vṛścika*, *dhana* (*dhanuḥ*), *makara*, *kubha* (*kumbha*), *mīna*. Each sign is further subdivided into 10 units of 3 degrees each and numbered as 3, 6, 9...24, 27, 30 in counter-clockwise direction. The divisions under Gemini (*mithuna*), however, are not numbered.

On this grid the positions of 24 prominent stars are marked on the basis of their longitudes (*dhruvaka*) and latitudes (*vikṣepa*)[33] and these points are connected to the nearest circles or bars in the form of star pointers (*nakṣatra-cañcu*), curved like tiger's claws, with the names of the respective stars engraved on them. Ten of these are outside the ecliptic, i.e. to the south of it; while fourteen are inside the ecliptic circle, i.e. to the north of the ecliptic. The stars to the south of the ecliptic have southern declinations; those to the north of the ecliptic have northern declinations.

The names on the star pointers are generally correct, but with occasional errors. Sometimes, the vocalic symbols (*mātrās*) on the top and below are missing. The engraver replaces Sanskrit *kha* with *ṣa* and *ba* with *va*. All the star names are taken from Mahendra Sūri's *Yantrarāja*.

Table 2 below lists the stars on the rete, starting from the vernal equinox and proceeding in the order of increasing longitude

The designation of no. 10 (Sirius, α Canis Maioris) as *ārdrālu* (short for *ārdrā-lubdhaka*) is problematic. In his *Yantrarāja*, Mahendra Sūri enumerates 32 astrolabe stars with their longitudes and latitudes.[34] The 14th star in the list is *ārdrā* (p. 26, verse 28). In the tables appended to the commentary, this star is mentioned as *ārdrā-lubdhaka* (pp. 37, 41) and the corresponding Arabic name is given as *sorāmānī* or *serāimānī*. This is clearly an incorrect transcription of *shi'rà yamānīyaḥ*, i.e. Sirius (α Canis Maioris).

Besides the present astrolabe, several other seventeenth century astrolabes employ Mahendra's designation for Sirius. Thus an astrolabe produced in 1644 for Maṇirāma[35] employs the star names together with their serial numbers as assigned by Mahendra Sūri. Here Sirius is labelled as '*ā 14*,' i.e., the 14th star in Mahendra's list; therefore *ā* stands for *ārdrā-lubdhaka*. In an astrolabe produced in 1618[36] and in another of 1669[37] Sirius is labelled as '*ā.lu.*,' i.e., short for *ārdrā-lubdhaka*. The Sanskrit University of Varanasi

[33]Mahendra Sūri, *Yantrarāja*, 1. 22-37; see also the tables in Malayendu Sūri's commentary on pp. 36-43.

[34]Mahendra Sūri, *Yantrarāja* 1.22-37.

[35]See n. 12 above.

[36]This astrolabe is with the Raja Dinkar Kelkar Museum, Pune (acc. no. 106 PMC).

[37]See n. 14 above.

S. No.	Star Name as engraved	Star Name full/correct	Identification	Popular Name
1	*samudrapakṣī*		β Ceti	Deneb Kaitos
2	*maghodara*	*Matsyodara*	β Andromedae	Mirach
3	illegible			
4	*pre*	*Pretaśirāḥ*	β Persei	Algol
5	*rohiṇī*		γ Tauri	Aldeberan
6	*ṣaṇmuṣa*	*Ṣaṇmukha*	α Aurigae	Capella
7	*mithunapādapa*	*Mithuna-pāda-dakṣiṇa*	κ Orionis	Saiph
8	*vamaskada*	*Mithuna-vāma-skandha*	γ Orionis	Bellatrix
9	*mithunaha*	*Mithuna-hasta*	α Orionis	Betelgeuse
10	*ardrālu*	*Ārdrā-lubdhaka*	α Canis Maioris	Sirius
11	*pra.vā*	*Prathama-bāla-śīrṣa*	α Geminorum	Castor
12	*lubdhakavadha*	*Lubdhaka-bandhu*	α Canis Minoris	Procyon
13	*maghā*		α Leonis	Regulus
14	*uttarāphā*	*Uttara-phālgunī*	β Leonis	Denebola
15	*kākāskanda*	*Kāka-skandha-pakṣa*	γ Corvi	Gienah
16	*citrā*		α Virginis	Spica
17	*svāti*		α Bootis	Arcturus
18	*viśāṣā*	*Viśākhā-mātṛ-maṇḍala*	α Coronae Borealis	Alphecca
19	*dhanukoṭi*	*Dhanuḥ-koti*	α Ophiuchi	Rasalhague
20	*dhanuśirā*	*Dhanuḥ-śarāgra*	$\mu^{1,2}$ Sagittarius	
21	*abhijit*		α Lyrae	Vega
22	*śravaṇa*		α Aquilae	Altair
23	*kubha*	*Kakudapuccha*	α Cygni	Deneb/Arided
24	*pū bhā*	*Pūrvabhādrapadā*	β Pegasi	Sheat

Table 2: Stars on the Astrolabe Rete

owns an astrolabe which is not dated, but can be assigned to the seventeenth century. Here Sirius is labelled as *ārdrā*.[38]

However, *ārdrā* and *lubdhaka* are generally treated as two distinctly separate stars, the former standing for Betelgeuse (α Orionis) and the latter for Sirius. A large majority of Sanskrit astrolabes follow this nomenclature.[39] It needs further investigation to explain this discrepancy in the use of the name *ārdrā* for astrolabe stars. All that can be said at this moment is that Sanskrit astrolabes of the seventeenth century seem to follow Mahendra Sūri in naming α Canis Maioris as *ārdrā* or *ārdrā-lubdhaka*, while the astrolabes of the later period treat *ārdrā* and *lubdhaka* as two separate stars.

Leaving aside the circular and horizontal bands and the star pointers, the rest of the plate is cut off, so that through the resulting gaps the readings on the latitude plate just below can be read off. In the present rete, this task of removing the inner spaces is performed rather crudely. Furthermore, the tips of five pointers (Nos. 1, 3, 11, 13, 20) are not properly cut out so that the pointers are clearly visible. While the workmanship of the latitude plates is inferior to that of the mater, the workmanship of the rete is much worse. The calligraphy of the letters and numerals is also different from that on the latitude plates and on the mater.

Mahendra Sūri[40] recommends that the first point of Capricorn (*makarāsya*, Arabic: *murī ra's al-Jady*) should be shaped as a projecting point so that it touches the scale on the rim as the rete is rotated around the centre. This feature does not occur in the present rete.

3 Authenticity and the Date of the Astrolabe

3.1 The Inscription

On the inner side of the mater is an inscription in six lines in elegant Devanāgarī letters of western Indian style (Figure 6), which reads as follows:

---

[38]See Sreeramula Rajeswara Sarma, 'Kaṭapayādi Notation on a Sanskrit Astrolabe' in: *The Archaic and Exotic*, op. cit., pp. 257-272.

[39]For example, all the five Sanskrit astrolabes in German collections designate α Orionis as *ārdrā* and α Canis Maioris as *lubdhaka*; cf. Sarma Sreeramula Rajeswara Sarma, "Indian Astronomical Instruments in German Collections," *XXX. Deutscher Orientalistentag, Freiburg, 24.-28. September 2007.* Ausgewählte Vorträge, hrsg. im Auftrag der DMG von Rainer Brunner et al. Online-Publikation, Februar 2008. http://orient.ruf.uni-freiburg.de/dotpub/sarma.pdf, Figs. 7, 9, 10, 11, 16.

[40]Mahendra Sūrī, *Yantrarāja*, 3.16, Appendix.

śrīgaṇādhipati<r> jayatu ||
svasti śrī saṃvat 1663 varṣe śāke 1528 prava
rttamāne māghavadi 1 pratipadātithau ravidine
amadāvādanagare mahāsuratrāṇa pātasāha śrī
salīmasāharājye yaṃtrarāja jo° caṃḍidāsaiṃ
karāvyu | *putra damodara paṭhanārthaṃ* || *śubhaṃ bhavatu* ||

Figure 6: The Inscription on the Inner Side of the Mater

May the lord of the *gaṇas* (= Gaṇeśa) be victorious.

May it be well. In Saṃvat 1663, Śaka 1528 current, on *pratipadā*, the first lunar day of the dark fortnight (*vadi*) of Māgha, on Sunday, at the city of Ahmadabad, during the reign of the Great Sultān, the Badshāh, the illustrious Salīm Shāh (i.e., Mughal Emperor Jahangir), [this] astrolabe (*yantrarāja*, lit. king of instruments) was caused to be made (*karāvyu*) by the astrologer Caṇḍidāsa for the purpose of the reading of [his] son Damodara. Let it be auspicious.

The sentence begins in Sanskrit, but ends in medieval Gujarati for *caṇḍidāsaiṃ karāvyu* is medieval Gujarati (for Sanskrit *caṇḍīdāsena kāritam*). However, such linguistic mixture is not unusual in the 'popular' Sanskrit in medieval Gujarat.[41] The engraver spelt the name Dāmodara wrongly as Damodara. The name of his father who commissioned the astrolabe should properly be Caṇḍīdāsa, but Caṇḍidāsa is not incorrect either. He has the title *jo°*, which is an abbreviation of Joṣī, 'astrologer'.

Now we come to the provenance of this astrolabe and the history of interpretation of the inscription. Several years ago it was said to be in the private collection of a certain Roberto Riva and to have been exhibited at the Museum of Natural History, Houston, Texas.[42] In 1976 it reached the auction house of Sotheby's who consulted Francis Maddison, the then curator of the Oxford Museum of the History of Science which holds perhaps the largest collection of astrolabes. On 9 November 1976 Maddison wrote to David Pingree, Professor of History of Mathematics at Brown University, requesting him to decipher the Sanskrit inscriptions. In this letter Maddison describes the astrolabe as 'a fairly crudely made Indian astrolabe, presumably of the 19th century, which I do not think should be categorized as a fake.' Unfortunately, with this sentence Maddison set the course of the history of the subsequent interpretations. David Pingree translated the inscription and the data on the six plates, transcribed the names of the 24 stars marked on the star map and identified them.

The Time Museum at Rockford, Illinois, acquired this astrolabe soon afterwards, probably from Sotheby's. An excellent catalogue of the astrolabes in this museum was prepared by Anthony Turner in 1985.[43] While preparing the catalogue, Turner once more consulted David Pingree and also Jean-Pierre Verdet of the Paris Observatory.

Pingree took objection to the expression *paṭhanārthaṃ*, 'for the sake of reading,' in the inscription: after all, an astrolabe is not read but used in observation. Lover of manuscripts as he is —he must have read thousands of

---

[41]Professor Nalini Balbir, Paris, has kindly provided me this information on medieval Gujarati and assured that such mixture is not unusual in seventeenth century Gujarat. For documents of such 'popular' Sanskrit of earlier centuries in Gujarat, see Ingo Strauch, *Die Lekhapaddhati-Lekhapañcāsikā: Briefe und Urkunden im mittelalterlichen Gujarat*, Berlin 2002.

[42]Christie's, South Kensington, *Time Measuring Instruments from THE TIME MUSEUM*, for sale by Auction, Thursday 14 April 1988, Catalogue, no. 157, pp. 98-99.

[43]A. J. Turner, *Astrolabes, Astrolabe-related Instruments*, The Time Museum, Catalogue of Collection, Vol. I, Part 1, Rockford 1985. The present astrolabe is illustrated and described on pp. 120-123 (No. 15).

manuscripts in several classical languages— he immediately concluded that the inscription was copied from a manuscript, in order to lend a veneer of antiquity to the astrolabe. Following his interpretation, Turner states in the Time Museum Catalogue:

> The inscription on the *mater* engraved in a different hand can hardly be admitted as evidence for the date of the instrument. Indeed Pingree has suggested that the inscription was taken from a manuscript treatise (as the phrase 'for the purpose of reading...' would suggest) whence it was inaccurately copied, giving rise to some minor errors and one unreadable word. If this suggestion be accepted, then one might hypothesize that the instrument was made at a relatively late date by a metal-worker who, having some knowledge of both Lahore astrolabes and of astrolabe literature, combined aspects of each with the Hindu tradition to produce this eclectic instrument.[44]

In 1988, the Time Museum decided to part with this astrolabe and several other instruments. Christie's held a special auction of these instruments on 14 April 1988 in London. The auction catalogue assigns the astrolabe to 'probably 19th century'. A footnote to the entry states:

> This astrolabe appears in most respects to belong to the nineteenth century tradition of astrolabes engraved in Sanskrit for use in the Hindu community. The inscription in the **mater** cannot reasonably be taken as the date of the astrolabe, but remains a puzzle. It has been suggested that it was copied from a manuscript treatise on the astrolabe.[45]

Therefore the auction house offered this astrolabe at a much lower price than it charged for a single plate Sanskrit astrolabe of the nineteenth century.[46] In this auction, it was acquired by Saul Moskowitz, a dealer in scientific instruments of Marblehead, Massachusetts, USA. After his death, his collections were disposed of and the present owner acquired the astrolabe in question.

In spite of the high esteem in which I hold Francis Maddison, David Pingree and Antony Turner, I have to differ from their view.[47]

---

[44]Ibid, p. 122.

[45]Christie's, South Kensington, *Time Measuring Instruments from THE TIME MUSEUM*, op. cit., Catalogue, no. 157, pp. 98-99.

[46]Ibid, no. 149, pp. 86-87.

[47]Francis Maddison's writings enriched my knowledge of the history of the astrolabe in different cultures. I had the privilege of reading Mahenda Sūri's *Yantrarāja* with David Pin-

When these evaluations were made in the 1980s, nothing much was known about Sanskrit astrolabes. Things improved considerably since then. In the course of my project, I have identified more than a dozen Sanskrit works which discuss the astrolabe. I have also located nearly one hundred Sanskrit astrolabes in various museums and private collections.

Pingree et al thought that early seventeenth century is too early a date for Sanskrit astrolabes. But as mentioned above, I found nine other Sanskrit astrolabes produced in this century, most of them in Gujarat, and published some of these. In the light of this material, we have to draw a different conclusion about the date of the present astrolabe.

Turner thought that the astrolabe was made by a 'metal-worker who, having some knowledge of both Lahore astrolabes and of astrolabe literature, combined aspects of each with the Hindu tradition to produce this eclectic instrument.' I have explained already how Sanskrit astrolabes are produced. I have said that unlike in Islam, making scientific instruments did not develop into a specialized profession among the Hindu metal-workers. The initiative for the production of Sanskrit astrolabes came first of all from the astronomer, who prepared the initial drawings, and coaxed some reluctant brass worker to prepare the astrolabe according to his drawings. That is why there are often orthographic errors in the star names.

Furthermore, it is not customary in India for the artisans to sign their products. But in Islamic culture, there are many instances of metal workers signing their creations.[48] Islamic astrolabists, who enjoyed great prestige, naturally signed their products with their names. In fact, even the earliest surviving astrolabe produced in the Islamic world in 927 bears the name of its maker Nasṭūlus.[49]

gree. It is from Anthony Turner's lucid writings that I learnt much of what I know today on the astrolabes. Moreover, all the three have helped me in various ways in my project of cataloguing Indian instruments.

[48]Cf. Kjeld von Folsach, *Islamic Art: The David Collection*, Copenhagen, 1990, p. 183: 'Except for the art of the book it is also in the field of metalwork that most artists' names were recorded, and in that of scientific instruments we nearly always know of the name of the maker.' The David Collection contains an astrolabe and a quadrant (p. 214, nos. 361 and 362), both of which are signed and dated by the respective makers, but more interesting is an ornate brass jug (p. 247, no. 347) which is inscribed with the name of Ali ibn Muhammad Ali Shahab al-Ghuri.

[49]David A. King, 'Early Islamic Astronomical Instruments in Kuwaiti Collections' in: Arlene Fullerton & Géza Fehérvári, *Kuwait: Arts and Architecture: A Collection of Essays*, Kuwait 1995, pp. 76-96.

Hindus, who were attempting to follow Muslims in the production of astrolabes, tried to emulate the custom of signing the products as well. It is reasonable to assume that the astrologer Caṇḍīdāsa commissioned the astrolabe for his son Dāmodara. He himself prepared the drawings and got them executed by an unnamed metal-worker. Caṇḍīdāsa also wanted to have the date and purpose of the instrument engraved on the astrolabe. In the Sanskritic culture of Hindus and Jains, such inscriptions occurred hitherto only in manuscripts. Therefore, Caṇḍīdāsa prepared the inscription for the astrolabe in the same style as was done in manuscripts. He did not *copy* the inscription from a manuscript as Pingree and Turner thought, but *imitated* the manuscript practice. Because Caṇḍīdāsa was consciously imitating the manuscript practice, he used the expression 'for the sake of reading' (*paṭhanārtham*).[50] One may also argue that Caṇḍīdāsa got this astrolabe made as a teaching tool for his son Dāmodara.[51] In any case, *paṭhanārtham* does not detract from the validity of the date in the early seventeenth century, especially in view of nine other Sanskrit astrolabes of this period. It has also been shown that what was thought to be an unreadable word (viz. *karāvyu*) is actually medieval Gujarati and that such linguistic mixtures were common in the popular Sanskrit employed in medieval Gujarat.

This is one of the first attempts, somewhat clumsy no doubt, of emulating the Islamic custom of engraving the date of manufacture and the name of the maker. Later the custom took deep roots in Sanskrit tradition and people began to versify the inscriptions as well.[52] It is worth noting that, while the inscriptions in Islamic astrolabes mention the maker of the astrolabe, those in Sanskrit astrolabes mention generally the patron who commissioned the

[50]What other expression would be more suitable for the astrolabe? The first reaction would be to say *vedhārdharthan*, 'for the sake of astronomical observation or measurement.' It is true that the astrolabe is used to sight a heavenly body and measure its altitude, but that is not the only function of the astrolabe; *gaṇanārthaṃ*, 'for the sake of computation' would not be entirely correct either because the astrolabe graphically shows the result that one would otherwise obtain by computation.

[51]This recalls the case of Geoffrey Chaucer who wrote his book on the astrolabe for his little son Louis. Cf. Geoffrey Chaucer, *The Treatise on the Astrolabe in: The Complete Works of Geoffrey Chaucer*, ed. Walter W. Skeat, vol. III, Oxford, impression of 1996.

[52]For example, a *Dhruvabhrama-yantra* made in 1785 (now in a private collection in Brussels) carries an inscription in *Indravajrā* metre:

munyabhravāraikamite śakābde śrī-kīrticandrasya nṛpādhipasya |
ājñānusārād akarot suyaṃtraṃ śrī-motilālābhidha-śilpisiṃhaḥ ||

'In the Śaka year 1707 (= AD 1785), following the orders of the lord of the kings, śrī Kīrticandra, śrī Motilāla, the lion among artisans, made this excellent instrument.'

astrolabe.[53]

The calligraphy of the numerals in the inscription matches perfectly with the numerals in the degree scales on the front and the back of the mater. The letters of the inscription are quite similar to those engraved on the top of the crown (Figure 7). Thus the numerals and letters in all the engravings on the mater, namely the short inscription on the crown, the long inscription on the inner side of the mater, the degree scales on the front and the back of the mater are all alike, and belong to the western Indian style of Devanāgarī of the early seventeenth century. Moreover, the graph of unequal hours in the upper right quadrant at the back indicates an early date.

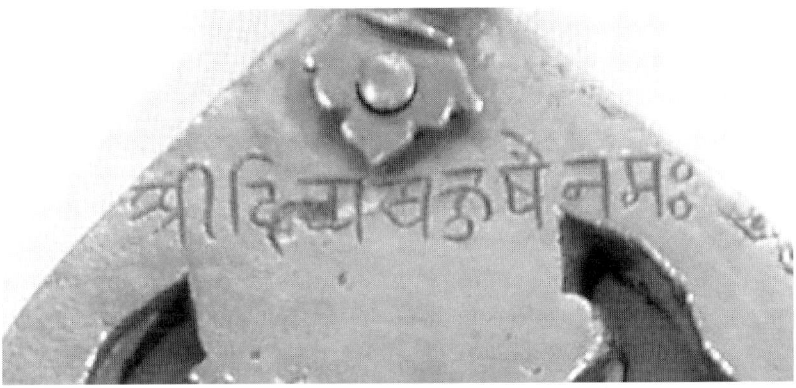

Figure 7: The Inscription on the Crown

The actual problem is not with the inscription on the mater, nor with the mater itself. The mater, a piece of excellent workmanship, and excellent calligraphy, in spite of minor errors in the inscription, fits very well in the early seventeenth century. The problem is with the remaining parts. For the numerals and letters on the six plates do not match with those on the mater; again the numerals and the letters on the rete do not match with those on the mater (Figure 8) or with those on the plates.

There is also a difference in the workmanship in these three components. The mater is elegantly produced. The plates do not show such good work-

[53]There are of course a few cases where the makers themselves signed their names on Sanskrit instruments. Thus Sonī Morārjī of Saurāṣṭra, Gujarat, inscribed his name on two identical *Dhruvabhrama-yantras* which he produced in 1815. Likewise Bhālūmal (*fl.* 1839-1850) of Lahore produced some twenty instruments of diverse types and inscribed them in Arabic/Persian or Sanskrit. Five of these carry Sanskrit legends and signatures in *Anuṣṭubh* metre.

manship, even though the almucantars lines are well engraved there. The rete displays very poor workmanship. Francis Maddison's observation that it is 'a fairly crudely made Indian astrolabe' applies especially to the rete. The rete is indeed very crude and does not match with the workmanship of the mater.

Figure 8: Above the Numerals on the Rim of the Mater; below the Numerals on the Rete

Clearly these three components, viz. mater (and the alidade with the sighting tube), plates and rete, are by three different hands and belong to three different periods. But the mention of the equinoctial shadow in addition to the longest day on the plates fits in well with the seventeenth century practice, as we have shown above. Likewise, the designation of Sirius as *ārdrā-lubdhaka* on the rete indicates an early period. The only conclusion that can emerge from this is that the plates and the rete were made at later periods to replace the damaged original components. The persons who commissioned these replacements did not design the plates and the rete anew, they had access to the original damaged parts which they got copied as best they could. Thus the plates and the rete, through prepared in later periods, retain some elements of the seventeenth century Sanskrit astrolabes. Therefore these replacements are also valuable as indirect witnesses to the seventeenth century practice.

3.2 The Date

The inscription reads 'Saṃvat 1663, Śaka 1528 current, on *pratipadā*, the first lunar day of the dark fortnight of Māgha, on Sunday'. Pingree converted this date to 1 February 1607. But when this date is reconverted using the

PANCANGA program,[54] we get Thursday Māgha śukla 5 Śaka 1528, instead of the original date. This happens because Pingree apparently did not pay attention to the expression *pravarttamāne* (current) in the inscription.

When we now convert Māgha kṛṣṇa 1 (*pūrṇimānta*) in the current Śaka year 1528 with the PANCANGA program, we get Monday, 26 December 1605. When this date is adjusted for the weekday, the result is Sunday, 25 December 1605. On that day the *pratipadā tithi* commences after sunrise.[55]

There is one more reason in support of the earlier date. Just two months previous to this date, i.e. on 24 October 1605 Salīm Shāh ascended the throne at Agra and assumed the name Nūr al-Dīn Jahāngīr. But the title is still new. Therefore the use of Salīm Shāh was more appropriate two months later on 25 December 1605 than fifteen months later on 1 February 1607.

Thus it cannot be doubted any more that the astrolabe was indeed produced originally in 1605. Until some other piece turns up, this *Yantrarāja* commissioned by the astrologer Caṇḍīdāsa for his son Dāmodara remains the earliest extant Sanskrit astrolabe. It is also worth noting that the original design of this astrolabe closely follows the prescriptions given in the earliest Sanskrit manual on the astrolabe, viz. the *Yantrarāja* by Mahendra Sūri.

Appendix

Mahendra Sūri's *Yantrarāja* on the Construction of the Astrolabe[56]

2.1-6: Constitution of the Astrolabe

ādau yantraṃ mṛṇmayaṃ dhātujaṃ vā vistīrṇaṃ ca svecchayā kārayitvā |
dairghyavyāsau pālivrittasya tasminn āryaiḥ kāryau yantracakrānumānāt ||1||
yāmye bhāge 'sya trikoṇaṃ kirīṭam īdṛg yantraṃ koṣṭhakāgāram uktam |
madhye tasya sveccayākṣāṃśakānāṃ patrāṇy anyāny unnatāṃśāśritāni ||2||
ekaṃ patraṃ connatāṃśasya patrād dvighnaṃ piṇḍe sādhanīyaṃ tato 'nyat |
laṅkotpannā rāśayo meṣamukhyāḥ saṃsthāpyante yatra dhiṣṇyaiḥ sametāḥ ||3||
patrāṇy evaṃ koṣṭhakāgāramadhye muktvā sādhyās teṣu pūrvādikāṣṭhāḥ |
pṛṣṭhe yantrasyāyate dve bhujāgre sūkṣme kṛtvā chidram antarbhujaṃ ca ||4||

[54] http://www.cc.kyoto-su.ac.jp/~yanom/pancanga/index.html created by M. YANO and M. FUSHIMI.

[55] Prof. Michio Yano of Kyoto Sangyo University, the co-author of the PANCANGA program, has kindly confirmed my conversion.

[56] The printed text (see n. 5 above) has many errors which are silently corrected here.

agre paścāt tasya cānte 'bdhikoṇe kḷptvā chidre yantranetre niveśye ।
chidre kḷptvā merukīlaṃ dalāliṃ tena kṣiptvā ghoṭikā saṃniveśyā ॥5॥
kṛtvā chidraṃ sūkṣmam asya trikoṇe tatra kṣiptvā kaṇṭakaṃ vṛścikābham ।
tasya prānte mudrikāṃ lambikākhyaṃ sūtraṃ deyaṃ yantraniṣpattir evam ॥6॥

First get an instrument (i.e. disc) (*yantra*) of desired size (*vistīrṇa*) prepared out of clay or of metal. Then the noble persons should fix on that [disc an upraised] rim (*pālivṛtta*) with height and breadth appropriate to the size of the astrolabe.

On the southern (i.e. the upper) part of the disc, [there should be] a triangular crown (*kirīṭa*). Such a component is called the mater (*koṣṭhakāgāra*, lit. store-house, repository). Inside this [are placed] as many latitude plates (*akṣāṃśakānāṃ patrāṇi*) as one wishes, which are endowed with [circles of equal] altitudes (*unnatāṃśa*).

Prepare another plate with double the thickness of the plates bearing the [circles of] altitudes. On this plate will be marked the zodiac signs starting from Aries, according to their rising times at the equator (*laṅkā*) together with [some prominent] fixed stars (*dhiṣṇya*).

Insert these plates inside the mater (*koṣṭhakāgāra*) and mark on [each face of] them the cardinal directions like east etc. At the back of the instrument, [attach] a long arm (i.e. alidade) with pointed tips and a hole at the middle.

At the front and back [of the arm] (i.e. on either end of the arm) set up (*niveśye*) two rectangular (*abdhikoṇa*) sights (*yantra-netra*), having made (*kḷptvā*) holes in them. Into the hole [at the middle of the alidade], insert the pin which represents the north celestial pole (*meru-kīla*) so that it passes through the series of latitude plates (*dala*). Into [the other end of] this pin, insert a horse-headed wedge (*ghoṭikā*).

In the triangle [of the crown], make a small hole at [the apex of] the triangular [crown], insert into it the shackle (*kaṇṭika*), which is shaped like the scorpion's sting, into it a ring (*mudrikā*) and then pass through it a string called the suspender (*lambika*). Thus is the astrolabe constituted.

3.1: Varieties of Astrolabes

yantraṃ proktaṃ ṣaḍvidhaṃ hy ānavatyā eka-dvi-trīṣv-aṅga-paṅkty-aṃśa-kḷptyā ।
dvedhāpy etat saumyayāmyaprabhedāt tanmiśratve miśrasaṃjñaṃ paraṃ ca ॥1॥

The astrolabe is said to be of six kinds [according as the altitude circles] are drawn (*kḷpti*, lit. arrangement) up to ninety [degrees], [with one circle] for each one (*eka*), two (*dvi*), three (*tri*), five (*iṣu*), six (*aṅga*), [or] ten (*paṅkti*) degrees (*aṃśa*). It is also of two varieties, northern (*saumya*) and southern (*yāmya*). When these two are combined, there is one more variety called the composite (*miśra*).

3.2-3: Back of the Astrolabe

vṛttadvayaṃ karkaṭakena pṛṣṭhe yantrasya nirmāya catur diśo 'ṅkyāḥ ǀ
prāgyāmyagāḥ koṣṭhagatās tataś ca sthāpyāḥ kha-nanda-pramitonnatāṃśāḥ ǁ2ǁ

vṛtte dvitīye likhitonnatāṃśā rekhā vilekhyā pratibhāgajātāḥ ǀ
pakṣatraye 'apakramajā vibhāgāḥ śaṅkuprabhā prāg gaditā tathā jyā ǁ3ǁ

On the back of the astrolabe, draw two annuli (*vṛtta*, lit. circle) with a pair of compasses (*karkaṭa*) and mark the four cardinal directions. Then between the east and west points, mark out ninety (*kha-nanda*) degrees of altitude in separate cells (*koṣṭha*).

In the second annulus, mark the lines for each degree corresponding to the degrees of altitude written [in the first annulus]. Then in three quadrants (*pakṣa*) draw respectively the units of declination (*apakramaja vibhāga*), gnomon shadows (*śaṅkuprabhā*) and the afore-mentioned sines (*jyā*).[57]

Thus the calibration of the back of the astrolabe (*pṛṣṭabhāgasādhana*) is complete.

3.4: Front of the Mater

atha yantre koṣṭhakāgārasya pūrvapakṣasādhanam āha ǀ
vṛttatraye pāligate kṛte prāg vṛtte kirīṭāntarato ghaṭīś ca ǀ
aṃśān abhīṣṭān kha-rasāgni-saṃkhyān rekhās tadīyās tadadho vilekhyāḥ ǁ4ǁ

[57]There is some confusion here in the specifications for the four quadrants. Both the text and the commentary appear to be mentioning only three quadrants. The commentary states that units of declination should be drawn in the south-western quadrant (*dakṣiṇapaścimāntarāla*), shadow squares in the north-western quadrant (*pascimasaumyāntarāla*) and the sine graph (*jīvāṅkāḥ*) in the north-eastern quadrant (*uttarapūrvāntarāla*). This makes sense only if is clearly stated that the shadows squares are drawn in the *two* lower quadrants. Moreover, the sine graph is usually drawn in the south-eastern quadrant, some times also in the south-western quadrant, but never in the north-eastern quadrant, as the commentary recommends.

Now he teaches the calibration of the front of the mater (*koṣṭhāgārasya pūrvapakṣa-sādhana*) of the astrolabe (*yantra*).

Having drawn three [concentric] circles on the rim (*pāli*), in the first circle, mark the *ghaṭīs*, starting from the middle of the crown (*kirīṭa*). [In the middle circle] mark 360 degrees at desired intervals. Below that, draw the lines of their subdivisions.

3.5-8: Latitude Plates

atha saumyayantre iṣṭākṣāṃśapatreṣūnnatavalayānāṃ sādhanam āha |
yantre ca saumye viracayya kaṃbāṃ patrānumānena vilikhya tatra |
bhāgān kharāmān racayed tadāṃśaiḥ karkādivṛttatritayaṃ daleṣu ||5||
digaṅkiteṣv eṣu ca madhyakendrād avācyarekhopari kendramānaiḥ |
cihne kṛte tacchirasaḥ pṛthutvamānena vṛttāni likhet sphuṭāni ||6||
bhūjākhyam ādyaṃ bhavatīha vṛttaṃ tataḥ paraṃ connatamaṇḍalāni |
teṣāṃ likhed ānavater vibhāgān śuddhān inādyunnatatāvagatyai ||7||
madhyāhnarekhām abhito 'sya karkavṛttasthite prāgapare vibhāge |
kramāl likhed bhāṃ paramaṃ dinaṃ taddeśābhidhānena tathākṣabhāgān ||8||

Now he teaches how to draw the altitude circles (*unnata-valaya*) in the desired latitude plates (*akṣāṃśa-patra*) in the northern astrolabe (*saumya-yantra*).

In the [case of the] northern astrolabe, prepare a ruler (*kambā*) according to the size of the plate (i.e. as long as the radius of the plate) and graduate it into 30 (*kha-rāma*) units (*bhāga*). With these units (*aṃśa*) draw the three circles of Cancer etc. upon the plates (*dala*).

On these plates, on which the cardinal directions have been marked [by means of N-S and E-W lines], make marks on the south line, starting from the centre, at distances measured by the values of eccentricity (*kendra*). Then from each of the marks as the centre draw clear circles with the measure of the radius (*pṛthutva*) [as given in the tables].

The first of such circles here is called the horizon (*bhūja*). Above that will be circles of altitude (*unnata-maṇḍala*). These [altitude circles] may be drawn clearly (*śuddha*) up to ninety degrees for determining the altitude (*unnatatā*) of the sun and other [celestial bodies].

On both sides, i.e., in the eastern and the western sides of the meridian or midday line (*madhyāhna-rekhā*) situated inside the tropic of Cancer, one should write successively the equinoctial shadow (*bhā*), maximum daylight (*paramadina*), the name of the locality (*deśābhidhāna*) and the degrees of latitude (*akṣa-bhāga*).

3.9: Unequal Hour Lines

atha saumyayantre horāsthāpanam āha |
kujād adho dvādaśadhā vibhajya mṛgādikarkāhvayamaṇḍaleṣu |
vidhāya vṛttāny abhitaḥ pratīcyā aṅkair niveśyā dvidaśāpi horāḥ ||

Now he teaches the construction of hour (*horā*) [lines] in the northern astrolabe.

Divide [the arcs of] the [three] circles of Capricorn, Aries and Cancer, which are situated below the horizon (*kuja*), into 12 parts each. Draw on both sides [of the meridian line, arcs of] circles (*vṛtta*) [through the respective points of division on these three circles] and number them from the west as the 12 hours (*horā*).

3.10-16: Rete

atha saumyayantre bhacakrapatre karkādivṛttatrayasādhanam | *bhacakra-kendrapramāṇena bhamaṇḍalasādhanaṃ tatra nirakṣameṣādilagnānām iṣṭa-bhāgānāṃ sthāpanam āha* |
bhacakrapatre 'pi puraiva kḷpte vṛttatraye 'bhyantarato 'ṣṭabhāgān |
hitvā kalāś cābdhiguṇān avācye vyāsārdhamānena vidhāya vṛttam ||10||
nirakṣameṣādivilagnamānān pūrvoditān prāci niveśya caindryāḥ |
vṛtte dvitīye tadadho niveśyās tadaṅkasaṃkhyā gaṇakair nijeṣṭāḥ ||11||
meṣe bhāgā kalāś cāpi dhiṣṇyāni gaganeṣavaḥ |
vṛṣe nandadṛśo 'bdhyakṣā yugme dantāḥ ṣaḍindavaḥ ||12||
vyutkramād eta eva syuḥ karkasiṃhakanīṣv api |
kanyādiṣaṭkamānaṃ syāt tulādau vaiparītyataḥ ||13||
vyāsākhyavṛtte 'jamukhāni santi lagnāni laṅkodayajāni yatra |
rekhā hy uḍūnāṃ pratibhāgajātās tatraiva kendrābhimukhā vilekhyāḥ ||14||
kendrāt pratīpaṃ viracayya cihnaṃ dyujyāpramāṇena ca karkaṭena |
tatraiva dhiṣṇyasya yathoditasya niveśyam asyāgram atīva sūkṣmam || 15||
dhanurmṛgāntar niśitaṃ vidheyaṃ cihnaṃ prasiddhaṃ makarāsyanāmnā |
yadbhrāmyamāṇaṃ gaṇakena viṣvaṅ muhur muhuś cumbati nāḍivṛttam ||16||
evaṃ saumyayantre bhacakrasādhanaṃ saṃpūrṇam ||

Now he teaches how to draw the three circles of Cancer and others on the rete (*bhacakrapatra*), how to draw the ecliptic circle (*bha-maṇḍala*) by means of the given value of the eccentricity of the ecliptic circle (*bhacakra-kendra*) and how to mark there the divisions of the zodiac signs according to their ascendants at zero degree latitude (*norakṣa-lagna*).

On the rete (*bhacakra-patra*) also, after having drawn the three circles [of Capricorn, Aries and Cancer] as before, leave out from the centre of the plate 8;34 units [which is the eccentricity of the ecliptic] in the south (*avācya*) (i.e. on the southern radius) [and with this point as the centre] draw a circle with the measure of the radius [of 21;26 units as shown in 1.66].

[On this circle], after having marked, from the east point onwards (*aindryāḥ*), the lengths of the right ascensions of the zodiac signs which have been mentioned before, in the second circle below the [previous one] the astronomer may mark their subdivisions according to his liking.

For Aries (*meṣa*) the degrees (*bhāga*) and minutes (*kalā*) [of right ascension] are 27/50; for Taurus (*vṛṣa*) 29/54 and for Gemini (*yugma*) 32/16.

The same in reverse order pertain to Cancer (*karka*), Leo (*siṃha*) and Virgo (*kanī*). The values of the six signs beginning with Virgo (*kanyā*) will apply in the reverse order to the six starting with Libra (*tulā*).

In the ecliptic circle (*vyāsākhya-vṛtta*) where there are the right ascensions (*laṅkodayajāni lagnāni*) of Aries and others, there draw the lines of stars, with the subdivisions, towards the centre.

With the measure of the day sine (*dyujyā*) make a mark away from the centre with the pair of compasses on that line. There at that point, create a very fine tip (*sūkṣmam agram*) for [pointer of] the afore-mentioned star.

Between Sagittarius (*dhanu*) and Capricorn (*mṛga*) affix the well-known mark called the *makarāsya* (face, or the first point of Capricorn), which touches the hour circle (*nāḍivṛtta*) again and again, when the rete is rotated all around by the astronomer.

Thus the preparation of the rete in the northern astrolabe is complete.

Indian planetary models: Āryabhaṭa to Nīlakaṇṭha[1]

K. Ramasubramanian[2]
IIT Bombay

Abstract

While the remarkable strides in mathematics made by the Kerala School[3], particularly in the field of analysis, has been fairly well recognized—at least among the scholarly circles, in the recent past—its significant contributions to the theoretical advances in astronomy has not received the attention it deserves. The present paper is an attempt to highlight the major revision in the traditional Indian planetary model that was brought out by Nīlakaṇṭha (1444-1545). It will also show, how the computational scheme for calculating the planetary longitudes and latitudes adopted by Nīlakaṇṭha is indeed a good approximation to the Keplerian model of planetary motion that was discovered more than 100 years later.

1 Introduction

The astronomers and mathematicians of Kerala, pioneered by the work of Mādhava (1340–1425) of Saṅgamagrāma took a "decisive step onward from

[1]This article, is actually a shorter version of a more detailed treatment of this topic, by K. Ramasubramanian, M. D. Srinivas and M. S. Sriram, which may be found in: *Tantrasaṅgraha* of Nīlakaṇṭha Somayājī, Tr. with detailed Explanatory Notes by K. Ramasubramanian and M. S. Sriram, Hindustan Book Agency, Delhi and Springer, London 2010, Appendix F: 'Traditional Indian planetary model and its revision by Nīlakaṇṭha Somayājī', pp. 487–535.

[2]e-mail: kramas@iitb.ac.in

[3]For the Kerala School of Astronomy, see for instance, K. V. Sarma, *A Bibliography of Kerala and Kerala-based Astronomy and Astrology*, Hoshiarpur 1972; K. V. Sarma, *A History of the Kerala School of Hindu Astronomy*, Hoshiarpur 1972.

the finite procedures of 'ancient' mathematics to treat their limit-passage to infinity which is the kernel of modern classical analysis".[4] By employing ingenious summation techniques, these mathematicians derived the infinite series for π, sine and cosine functions and also developed fast convergent approximations to them.[5] Besides these remarkable contributions in mathematics, how the Kerala school also made equally significant discoveries in astronomy, particularly planetary theory, would be discussed in the present article.

Mādhava's disciple Parameśvara of Vaṭasseri (c. 1380-1460)—the originator of *Dṛg-gaṇita* system, which replaced the older *Parahita* system—is reputed to have made continuous and careful observations for over fifty-five years before introducing the new system. He also discussed the geometrical picture of planetary motion as would follow from the traditional Indian planetary model in his brilliant commentary on *Āryabhaṭīya*, as well as some of his other works. Nīlakaṇṭha Somayājī of Tṛkkaṇṭiyur (c. 1444-1545), a disciple of Parameśvara's son Dāmodara, carried out a fundamental revision of the traditional planetary theory. In his treatise *Tantrasaṅgraha*, composed in 1500, Nīlakaṇṭha outlines the detailed computational scheme of his revised planetary model. For the first time in the history of astronomy, Nīlakaṇṭha proposed that in the case of an interior planet (Mercury or Venus), the *manda*-correction or the equation of centre should be applied to what was traditionally identified as the *śighrocca* of the planet—which, in the case of the interior planets, corresponds to what we currently refer to as the mean heliocentric planet. This was a radical departure from the traditional Indian planetary model where the *manda*-correction for an interior

[4]C. T. Rajagopal and M. S. Rangachari (June 1978). 'On an untapped source of medieval Keralese mathematics', *Archive for History of Exact Sciences*, 18 (2), p. 101.

[5]For overviews of the Kerala tradition of mathematics, see S. Parameswaran, *The Golden Age of Indian Mathematics*, Kochi 1998; G. G. Joseph, *The Crest of the Peacock: Non-European Roots of Mathematics*, 2nd Ed. Princeton 2000; C. K. Raju, *Cultural Foundations of Mathematics: The Nature of Mathematical Proof and the Transmission of the Calculus from India to Europe in the 16th c. C. E.*, Pearson Education, Delhi 2007. Kim Plofker, *Mathematics in India: From 500* BCE *to 1800* CE, Princeton 2009. See also a review of this book by David Mumford, *Notices of AMS*, March 2010. To have a flavor of a original source work, and the nature of argument adduced in it in order to justify a certain result, see: *Gaṇita-yukti-bhāṣā* (Rationales in Mathematical Astronomy) of Jyeṣṭhadeva, Ed. and Tr. by K. V. Sarma, with Explanatory Notes by K. Ramasubramanian, M. D. Srinivas and M. S. Sriram, 2 Vols, Hindustan Book Agency, New Delhi 2008. Reprint Springer 2009. (Editor's note: for the treatment of π, sine and cosine functions by Kerala mathematicians, see already the articles by J.M.Delire above and by K.Mahesh, Venketeswara Pai R. and K.Ramasubramanian below.)

planet was applied to the mean Sun.[6]

In his *Āryabhaṭīya-bhāṣya*, Nīlakaṇṭha explains the rationale behind his revision of the traditional planetary theory. This has to do with the fact (which was noticed by several Indian astronomers prior to Nīlakaṇṭha) that the traditional Indian planetary model employed entirely different schemes for computing the latitudes of the exterior and the interior planets. While the latitudes of the exterior planets were computed from their so called *manda-sphuṭa* (which corresponds to what we currently refer to as the true heliocentric planet), the latitudes of the interior planets were computed from their so called *śīghrocca*. Nīlakaṇṭha argued that since the latitude should be dependent on the deflection (from the ecliptic) of the planet itself and not of any other body, what was traditionally referred to as the *śīghrocca* of an interior planet should be identified with the planet itself. Nīlakaṇṭha also showed that this would lead to a unified treatment of the latitudinal motion of all the planets—interior as well as exterior.[7]

The geometrical picture of planetary motion that can be deduced from the computational scheme has been described by Nīlakaṇṭha in his *Āryabhaṭīya-bhāṣya*, as well as *Golasāra* and *Siddhānta-darpaṇa*. In a remarkable short tract called *Grahasphuṭānayane vikṣepavāsanā*, which seems to be written after the *Āryabhaṭīya-bhāṣya* as the former quotes extensively from the latter, Nīlakaṇṭha succinctly describes his cosmological model—according to which the five planets, Mercury, Venus, Mars, Jupiter and Saturn, go around the mean Sun in an eccentric orbit (inclined to the ecliptic), while the mean Sun itself goes around the Earth.[8] Following this, Nīlakaṇṭha also states that the dimensions of *śīghra* epicycles are specified by measuring the orbit of the mean Sun around the Earth in terms of the planetary orbit in the case of the exterior planets, and they are specified by measuring the planetary orbit (which is smaller) in terms of the orbit of the mean Sun in the case of the interior planets. This remarkable relation[9] follows clearly from the identifi-

[6]It had also been a general feature of all ancient planetary theories in the Greco-European and the Islamic traditions of astronomy, till the work of Kepler, that the equation of centre for an interior planet was wrongly applied to the mean Sun.

[7]In fact, it has been noted in a later text, *Vikṣepagolavāsanā*, that Nīlakaṇṭha pioneered a revision of the traditional planetary theory in order to arrive at a unified formulation of the motion in latitude of both the interior and exterior planets.

[8]This cosmological model is the same as the one proposed by Tycho Brahe, albeit on entirely different considerations, towards the end of sixteenth century.

[9]The *śīghra* epicycle is essentially the same as the epicycle associated with so called "solar anomaly" in the Greco-European tradition of astronomy, and the above relation is the same as the one proposed by Nicholas Copernicus (around the same time as Nīlakaṇṭha) by

cation of the *śighrocca* of all the planets with physical mean Sun, a fact also stated by Nīlakaṇṭha in his *Āryabhaṭīya-bhāṣya*.

Before taking up the various aspects of the revised planetary model of Nīlakaṇṭha it is essential to understand the traditional Indian planetary model, which had been in vogue at least from the time of Āryabhaṭa (c. 499). We shall therefore devote the initial sections of this paper to a detailed exposition of the traditional Indian planetary theory and important developments in it prior to the work of Nīlakaṇṭha.

2 The traditional Indian planetary model

In the Indian astronomical tradition, at least from the time of Āryabhaṭa (499 CE), the procedure for calculating the geocentric longitudes of the planets consists essentially of two steps:[10] First, the computation of the mean longitude of the planet known as *madhyama-graha*, and second, the computation of the true or observed longitude of the planet known as *sphuṭa-graha*.

The mean longitude is calculated for the desired day by computing the number of mean civil days elapsed since the epoch (this number is called the *ahargaṇa*) and multiplying it by the mean daily motion of the planet. Having obtained the mean longitude, a correction known as *manda-saṃskāra* or manda correction is applied to it. In essence, this correction takes care of the eccentricity of the planetary orbit around the Sun. The equivalent of this correction is termed equation of centre in modern astronomy, and is a consequence of the elliptical nature of the orbit. The longitude of the planet obtained by applying the *manda*-correction is known as *manda-sphuṭa-graha* or simply *manda-sphuṭa*.

While *manda-saṃskāra* is the only correction that needs to be applied in the case of the Sun and the Moon, for obtaining their true longitudes (*sphuṭa-grahas*), in the case of the other five planets two corrections, namely *manda-saṃskāra* and *śīghra-saṃskāra*, are to be applied to the mean longitude in order to obtain their true longitudes. Here again, we divide the five planets

identifying this epicycle as the orbit of the Earth around the Sun in the case of the exterior planets and as the orbit of the planet itself in the case of the interior planets.

[10] For a general review of Indian astronomy, see D. A. Somayājī, *A Critical Study of Ancient Hindu Astronomy*, Dharwar 1972; S. N. Sen and K. S. Shukla (eds), *A History of Indian Astronomy*, New Delhi 1985 (Rev. Edn. 2000); B. V. Subbarayappa, and K. V. Sarma (eds.), *Indian Astronomy: A Source Book*, Bombay 1985; S. Balachandra Rao, *Indian Astronomy: An Introduction*, Hyderabad 2000; B. V. Subbarayappa, *Tradition of Astronomy in India: Jyotiḥśāstra*, PHISPC Vol. IV, Part 4, Centre for Studies in Civilizations, New Delhi 2008.

into two groups—the interior, comprising of Mercury and Venus, and the exterior, comprising of Mars, Jupiter and Saturn—not necessarily for the purpose of convenience in discussion but also because of the fact that they are treated differently while applying these corrections.

The *śīghra-saṃskāra* is applied to the *manda-sphuṭa-graha* to obtain the true geocentric longitude known as *sphuṭa-graha*. As would be seen later, the *śīghra* correction essentially converts the heliocentric longitude into geocentric longitude. We shall now briefly discuss the details of the *manda-saṃskāra* which will be followed by a discussion on *śīghra-saṃskāra* for the exterior and the interior planets respectively.

2.1 *Manda-saṃskāra* : Epicyclic and eccentric models

As mentioned earlier, the *manda-saṃskāra* essentially accounts for the eccentricity of the planetary orbit. This may be explained with the help of Figure 1. Here, O is the centre of the *kakṣyāmaṇḍala*[11] on which the mean planet P_0 is assumed to be moving with mean uniform velocity. $O\Gamma$ is the reference line usually chosen to be the direction of *Meṣādi*. The *kakṣyāmaṇḍala* is taken to be of radius R known as *trijyā*.[12] The longitude of the mean planet P_0 moving on this circle is given by

$$\Gamma\hat{O}P_0 = \textit{madhyama-graha} = \theta_0. \tag{1}$$

The longitude of the *manda-sphuṭa-graha* P given by $\Gamma\hat{O}P$ is to be obtained from θ_0, and this can be obtained either by an eccentric or epicyclic model. The procedure for obtaining the longitude of the *manda-sphuṭa-graha* by either of the two models involves the longitude of *mandocca*. In Figure 1, OU represents the direction *mandocca* whose longitude is given by

$$\Gamma\hat{O}U = \textit{mandocca} = \theta_m. \tag{2}$$

The modern equivalent of *mandocca* is *apoapsis*—apogee in the case of the Sun and the Moon and aphelion in the case of the five planets. Around the

[11] The centre of the *kakṣyāmaṇḍala* is generally referred to as *bhagola-madhya* (centre of the celestial sphere), and it coincides with the centre of the Earth in the case of the Sun and the Moon.

[12] The value of *trijyā* is chosen such that one minute of arc in the circle corresponds to unit length. This implies that $2\pi R = 21600$ or $R \approx 3437.74$, which is taken to be 3438 in most of the Indian texts. (Editor's note: *trijyā* refers to the rsine (*jyā*) of three (*tri*) *rāśi* (=30°), which equals the radius.)

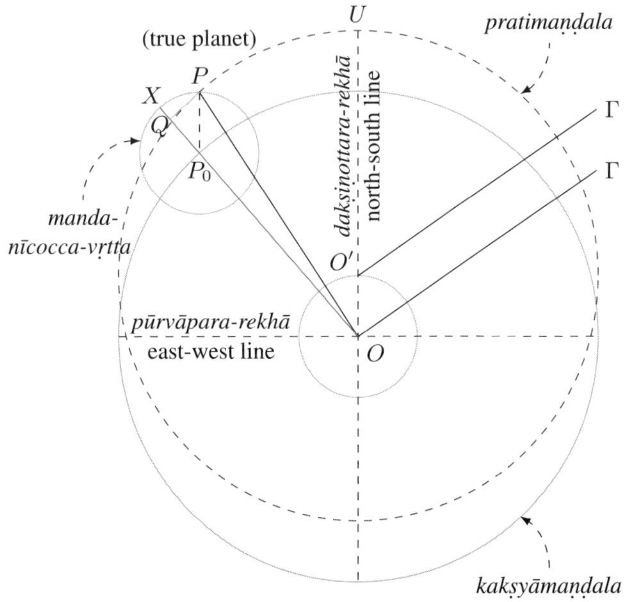

Figure 1: The epicyclic and eccentric models of planetary motion.

mean planet P_0, a circle of radius r is to be drawn. This circle is known as the *manda-nīcocca-vṛtta*[13] or simply as *manda-vṛtta* (epicycle). The texts specify the value of the radius of this circle r $(r \ll R)$, in an appropriate measure, for each planet.

At any given instant of time, the *manda-sphuṭa-graha* P is to be located on this *manda-nīcocca-vṛtta* by drawing a line from P_0 along the direction of *mandocca* (parallel to OU). The point of intersection of this line with the *manda-nīcocca-vṛtta* gives the location of the planet P. Since this method of locating the *manda-sphuṭa-graha* involves the construction of an epicycle around the mean planet, it is known as the epicyclic model.

Alternatively, one could draw the *manda-nīcocca-vṛtta* of radius r centered around O, which intersects OU at O'. With O' as centre, a circle of radius R (shown by dashed lines in the figure) is drawn. This is known as *pratimaṇḍala* or the eccentric circle. Since P_0P and OO' are equal to r, and

[13]The adjective *nīcocca* is given to this *vṛtta* (circle) because, in this conception, it moves from *ucca* (above) to *nīca* (below) on the deferent circle along with the mean planet P_0. The other adjective *manda* is to suggest that this circle plays a crucial role in the explanation of the *manda-saṃskāra*.

they are parallel to each other, $O'P = OP_0 = R$. Hence, P lies on the eccentric circle. Also,

$$\Gamma\hat{O}'P = \Gamma\hat{O}P_0 = \textit{madhyama-graha} = \theta_0. \tag{3}$$

Thus, the *manda-sphuṭa-graha* P can be located on an eccentric circle of radius R centered at O' (which is located at a distance r from O in the direction of *mandocca*), simply by marking a point P on it such that $\Gamma\hat{O}'P$ corresponds to the the mean longitude of the planet. Since this process involves only an eccentric circle, without making a reference to the epicycle, it is known as eccentric model. Clearly, the two models are equivalent to each other.

2.2 Calculation of *manda-sphuṭa*

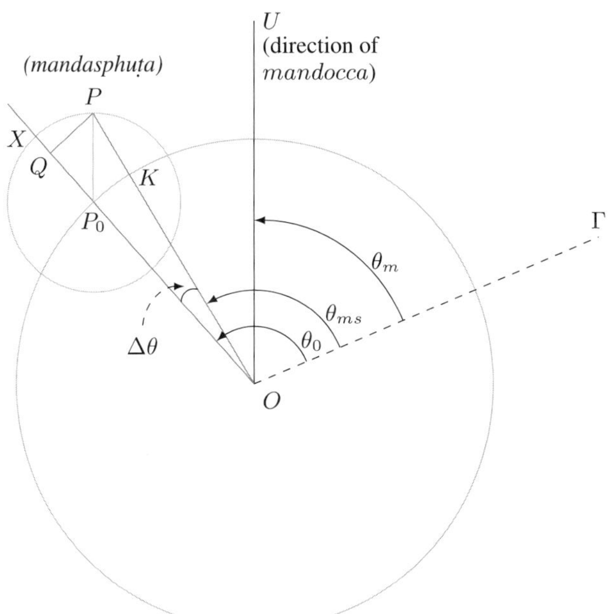

Figure 2: Geometrical construction underlying the rule for obtaining the *manda-sphuṭa* from the *madhyama* using the epicycle approach.

The formula presented by the Indian astronomical texts for the calculation of *manda-sphuṭa*—the longitude of the planet obtained by applying the *manda-saṃskāra* (equation of centre) to the mean longitude of the planet—and the underlying geometrical picture can be understood with the help of Figure 2.[14] Here, $\theta_{ms} = \Gamma\hat{O}P$ represents the *manda-sphuṭa* which is to be determined from the position of the mean planet (*madhyama-graha*) P_0. Clearly,

$$\begin{aligned} \theta_{ms} &= \Gamma\hat{O}P \\ &= \Gamma\hat{O}P_0 - P\hat{O}P_0 \\ &= \theta_0 - \Delta\theta. \end{aligned} \tag{4}$$

Since the mean longitude of the planet θ_0 is known, the *manda-sphuṭa* θ_{ms} is obtained by simply subtracting $\Delta\theta$, from the *madhyama*. The expression for $\Delta\theta$ can be obtained by making the following geometrical construction. We extend the line OP_0, which is the line joining the centre of the *kakṣyāmaṇḍala* and the mean planet, to meet the epicycle at X. From P drop the perpendicular PQ on OX. Then,

$$\begin{aligned} U\hat{O}P_0 &= \Gamma\hat{O}P_0 - \Gamma\hat{O}U \\ &= \theta_0 - \theta_m, \end{aligned} \tag{5}$$

is the *manda-kendra* (*madhyama - mandocca*), whose magnitude determines the magnitude of $\Delta\theta$ (see (8) below). Also, since P_0P is parallel to OU (by construction), $P\hat{P_0}Q = U\hat{O}P_0 = (\theta_0 - \theta_m)$. Hence, $PQ = r\sin(\theta_0 - \theta_m)$ and $P_0Q = r\cos(\theta_0 - \theta_m)$. Since the triangle OPQ is a right angled at Q, the hypotenuse $OP = K$ (known as *manda-karṇa*) is given by

$$\begin{aligned} K = OP &= \sqrt{OQ^2 + QP^2} \\ &= \sqrt{(OP_0 + P_0Q)^2 + QP^2} \\ &= \sqrt{\{R + r\cos(\theta_0 - \theta_m)\}^2 + r^2\sin^2(\theta_0 - \theta_m)}. \end{aligned} \tag{6}$$

Again from the triangle POQ, we have

$$\begin{aligned} K\sin\Delta\theta &= PQ \\ &= r\sin(\theta_0 - \theta_m). \end{aligned} \tag{7}$$

Multiplying the above by R and dividing by K we have,

$$R\sin\Delta\theta = \frac{r}{K}\,R\sin(\theta_0 - \theta_m). \tag{8}$$

[14] It may be noted that Figure 2 is the same as Figure 1, with certain circles and markings removed from the latter and certain others introduced in the former for the purposes of clarity.

In the Āryabhaṭan school, the radius of *manda*-epicycle is assumed to vary in the same way as the *karṇa*, as explained for instance, by Bhāskara I (c. 629) in his *Āryabhaṭīya-bhāṣya*, and also in his *Mahābhāskarīya*. Thus the relation (8) reduces to

$$R \sin \Delta\theta \quad = \quad \frac{r_0}{R}\, R \sin(\theta_0 - \theta_m), \tag{9}$$

where r_0 is the mean or tabulated value of the radius of *manda*-epicycle.

2.3 *Aviśiṣṭa-manda-karṇa*: Iterated hypotenuse

According to geometrical picture of planetary motion given by Bhāskara I, the radius of the epicycle *manda-nīcocca-vṛtta* (r) employed in the the *manda*-process, is not a constant. It keeps continuously varying in consonance with the hypotenuse, *manda-karṇa* (K) in such a way that their ratio is always maintained constant and is equal to the ratio of the mean epicycle radius (r_0)—whose value is specified in the texts—to the radius of the deferent circle (R). Thus, according to Bhāskara, as far as the *manda*-process is concerned, the motion of the planet on the epicycle is such that the following equation is always satisfied.

$$\frac{r}{K} = \frac{r_0}{R}. \tag{10}$$

If this were the case, then the question arises as to how one could obtain the *manda-karṇa* as well as the the radius of *manda-nīcocca-vṛtta* at any given instant. For this Bhāskara provides an iterative procedure called *asakṛt-karma*, by which both r and K are simultaneously obtained. We explain this with the help of Figure 3*a*. Here P_0 represents the mean planet around which an epicycle of radius r_0 is drawn. The point P_1 on the epicycle is chosen such that PP_1 is parallel to the direction of *mandocca* OU.

Now, the first hypotenuse (*sakṛt-karṇa*) is found from r_0 using the relation

$$OP_1 = K_1 = [(R \sin(\theta_0 - \theta_m))^2 + (R \cos(\theta_0 - \theta_m) + r_0)^2]^{\frac{1}{2}}. \tag{11}$$

From K_1, using (10), we get the next approximation to the radius $r_1 = \frac{r_0}{R} K_1$, and the process is repeated. From r_1 we get the next approximation to the *karṇa*

$$K_2 = [\{R \sin(\theta_0 - \theta_m)\}^2 + \{R \cos(\theta_0 - \theta_m) + r_1\}^2]^{\frac{1}{2}}. \tag{12}$$

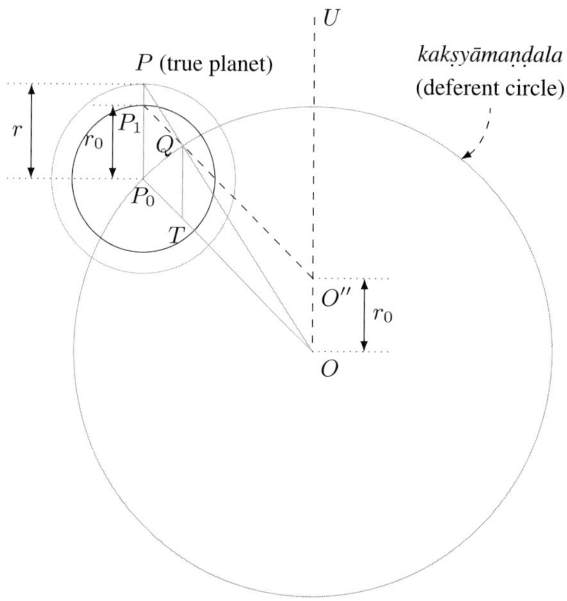

Figure 3*a*: The variation of the radius of the *manda* epicycle with the *manda-karṇa*.

and from that we get $r_2 = \frac{r_0}{R} K_2$ and so on, till the radii and the *karṇas* do not change (*aviśeṣa*). The term *aviśeṣa* means not distinct. In the present context it means that the successive *karṇas* are not distinct from each other. That is, $K_{i+1} \approx K_i = K$. If this is satisfied, then $r_{i+1} \approx r_i = r$. Consequently, the equation giving the *manda*-correction (8) becomes

$$R \sin \Delta\theta \quad = \quad \frac{r}{K} R \sin(\theta_0 - \theta_m) = \frac{r_0}{R} R \sin(\theta_0 - \theta_m). \qquad (13)$$

Thus the computation of *mandaphala* involves only the mean epicycle radius and the value of *trijyā*. It does not involve the value of *manda-karṇa*. It can be shown that the iterated *manda-karṇa* is actually given (in the limit) by OP in Figure 3*a*, where the point P is obtained as follows.[15] Consider a point O'' at a distance of r_0 from O along the direction of *mandocca* OU and draw $O''P_1$ so that it meets the concentric[16] at Q. Then produce OQ to meet the extension of P_0P_1 at P.

[15]See for instance, the discussion in *Mahābhāskarīya*, Ed. and Tr. by K. S. Shukla, Lucknow 1960, pp. 111-19.

[16]Editor's note: deferent or *kakṣyā maṇḍala*

Mādhava of Saṅgamagrāma, the renowned mathematician and astronomer of 14th century, by carefully analysing the geometry of the problem, came up with a brilliant method to find the *aviśiṣṭa-mandakarṇa* without doing an iterative process. This method due to Mādhava is explained in the following section.

2.4 Mādhava's formula for *aviśiṣṭa-mandakarṇa*

Mādhava's procedure for determining the *aviśiṣṭa-mandakarṇa* involves finding a new quantity called *viparyayakarṇa* or *viparītakarṇa*. The term *viparītakarṇa* literally meaning 'inverse-hypotenuse', and is nothing but the radius of the *kakṣyāvṛtta* when the *mandakarṇa* is taken to be *trijyā*, R. The following verses from *Tantrasaṅgraha* present the way of obtaining the *aviśiṣṭa-mandakarṇa* proposed by Mādhava that circumvents the iterative process.

> विस्तृतिदलदोःफलकृतिवियुतिपदं कोटिफलविहीनयुतम्।
> केन्द्रे मृगकर्किगते स खलु विपर्ययकृतो भवेत् कर्णः ॥
> तेन हृता त्रिज्याकृतिः अयत्नविहितोऽविशेषकर्णः स्यात् ।
>
> The square of *doḥphala* is subtracted from the square of *trijyā* and its square root is taken. The *koṭiphala* is added to or subtracted from this depending upon whether the *kendra* (anomaly) is within six signs beginning from *Karki* (Cancer) or *Mṛga* (Capricorn). This gives the *viparyayakarṇa*. The square of *trijyā* divided by this *viparyayakarṇa* is the *aviśeṣakarṇa* (iterated-hypotenuse) obtained without any effort [of iteration].

The rationale behind the formula given for *viparītakarṇa* is outlined in the Malayalam text *Yuktibhāṣā*, and can be understood with the help of Figures $3a$ and $3b$. In these figures P_0 and P represent the mean and the true planet respectively. N denotes the foot of perpendicular drawn from the true planet P to the line joining the centre of the circle and the mean planet. NP is equal to *doḥphala*. Let the radius of the *karṇavṛtta* OP be set equal to *trijyā* R. Then the radius of the *uccanīcavṛtta* P_0P is r_0, as it is in the measure of the *karṇavṛtta*. In this measure, the radius of the *kakṣyāvṛtta* $OP_0 = R_v$, the *viparītakarṇa* and is given by

$$\begin{aligned} R_v &= ON \pm P_0N \\ &= \sqrt{R^2 - (r_0 \sin(\theta_0 - \theta_m))^2} \pm |r_0 \cos(\theta_0 - \theta_m)|. \qquad (14a) \end{aligned}$$

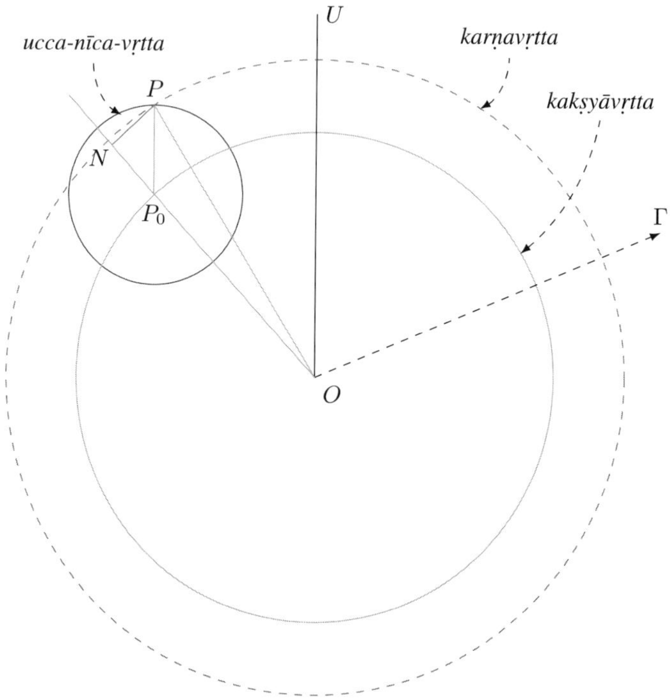

Figure 3*b*: Determination of *viparītakarṇa* when the *kendra* is in the first quadrant.

Nīlakaṇṭha has also given another alternative expression for the *viparītakarṇa* in terms of the longitude θ_{ms} of the *manda-sphuṭa*.

$$R_v \quad = \quad \sqrt{R^2 + r_0^2 - 2Rr_0\cos(\theta_{ms} - \theta_m)}. \qquad (14b)$$

In Figure 3*a*, Q is a point where $O''P_1$ meets the concentric. OQ is produced to meet the extension of P_0P_1 at P. Let T be the point on OP_0 such that QT is parallel to P_0P_1. Then it can be shown that $OT = R_v$ is the *viparītakarṇa*. Now, in triangle OQT, $OQ = R$, $QT = P_1P_0 = r_0$ and $O\hat{Q}T = P\hat{O}U = (\theta_{ms} - \theta_m)$ and we have

$$OT = \sqrt{R^2 + r_0^2 - 2Rr_0\cos(\theta_{ms} - \theta_m)} = R_v. \qquad (14c)$$

Now, since triangles OQT and OPP_0 are similar, we have

$$\frac{OP}{OP_0} = \frac{OQ}{OT} \quad = \quad \frac{R}{R_v}$$

$$\text{or,}\quad OP = K = \frac{R^2}{R_v}. \tag{15}$$

Thus we have obtained an expression for *aviśiṣṭa-mandakarṇa* in terms of *trijyā* and *viparītakarṇa*. As the computation of *viparītakarṇa* as given by (14a) does not involve iteration, the *aviśiṣṭa-mandakarṇa* can be obtained in one stroke using (15) without having to go through the arduous iterative process.

2.5 *Manda-saṃskāra* for exterior planets

We shall now discuss the details of *manda* correction for the case of the exterior planets, namely Mars, Jupiter and Saturn as outlined in the traditional texts of Indian astronomy. The texts usually specify the number of revolutions (*bhagaṇas*) made by the planets in a large period known as *Mahāyuga*. In Table 1, we list the *bhagaṇas* as specified in the texts *Āryabhaṭīya* and *Tantrasaṅgraha*. In the same table, we have also given the corresponding sidereal period of the planet in civil days along with the modern values for the same.

In the case of exterior planets, while the planets move around the Sun they also move around the Earth, and consequently, the mean heliocentric sidereal period of the planet is the same as the mean geocentric sidereal period. Therefore, the *madhyama-graha* or the mean longitude of the planet, as obtained from the above *bhagaṇas*, would be the same as the mean heliocentric longitude of the planet as we understand today. Now the *manda-saṃskāra* is applied to *madhyama-graha* to obtain the *manda-sphuṭa-graha*. As we shall see below, this *manda* correction is essentially the same as the equation of centre in modern astronomy and thus the *manda-sphuṭa-graha* would essentially be the true heliocentric longitude of the planet.

It was shown above (9) that the magnitude of the correction $\Delta\theta$ to be applied to the mean longitude is given by

$$R\sin\Delta\theta = \frac{r_0}{R}\,R\sin(\theta_0 - \theta_m), \tag{16}$$

If $\frac{r_0}{R}$ is small in the above expression, then $\sin\Delta\theta \ll 1$ and we can approximate $\sin\Delta\theta \approx \Delta\theta$. Hence, (16) reduces to

$$\Delta\theta = \frac{r_0}{R}\sin(\theta_0 - \theta_m). \tag{17}$$

Planet	Revolutions (in *Āryabhaṭīya*)	Sidereal Period (in *Āryabhaṭīya*)	Revolutions (in *Tantrasaṅgraha*)	Sidereal Period (in *Tantrasaṅgraha*)	Modern Period
Sun	43,20,000	365.25868	43,20,000	365.25868	365.25636
Moon	5,77,53,336	27.32167	5,77,53,320	27.32168	27.32166
Moon's apogee	4,88,219	3231.98708	4,88,122	3232.62934	3232.37543
Moon's node	2,32,226	6794.74951	2,32,300	6792.58502	6793.39108
Mercury's *śīghrocca*	1,79,37,020	87.96988	1,79,37,048	87.96974	87.96930
Venus's *śīghrocca*	70,22,288	224.69814	70,22,268	224.70198	224.70080
Mars	22,96,824	686.99974	22,96,864	686.98778	686.97970
Jupiter	3,64,224	4332.27217	3,64,180	4332.79559	4332.58870
Saturn	1,46,564	10766.06465	1,46,612	10762.53990	10759.20100

Table 1: The *bhagaṇas* and sidereal period of the planets.

As $\Delta\theta = \theta_0 - \theta_{ms}$, in this approximation, we have

$$\theta_{ms} \approx \theta_0 - \frac{r_0}{R}\sin(\theta_0 - \theta_m). \tag{18}$$

In the Keplerean picture of planetary motion, the equation of centre to be applied to the mean heliocentric longitude of the planet is given—to the first order in eccentricity—by the equation

$$\Delta\theta \approx (2e)\sin(\theta_0 - \theta_m). \tag{19}$$

Now, comparing (19) and (17), we see that the *manda* correction closely approximates the equation of centre as understood in modern astronomy if the values of $\frac{r_0}{R}$ are fairly close to $2e$.

The values of $\frac{r_0}{R}$ for different planets as specified in *Āryabhaṭīya* and *Tantrasaṅgraha* are listed in Table 2. It may be noted here that the ratios specified in the texts are close to twice the value of the eccentricity ($2e$) associated with the planetary orbits. In Table 2, the modern values of $2e$ are listed following the textbook by Smart.[17]

2.6 *Manda-saṃskāra* for interior planets

For the interior planets Mercury and Venus, since the mean geocentric sidereal period of the planet is the same as that of the Sun, the ancient Indian astronomers took the mean Sun as the *madhyama-graha* or the mean planet. Having taken the mean Sun as the mean planet, they also prescribed the application of *manda* correction or the equation of centre characteristic of the planet, to the mean Sun, instead of the mean heliocentric planet. Therefore, the *manda-sphuṭa-graha* in the case of an interior planet, as computed from (17) in the traditional planetary model, is just the mean Sun, with a correction applied, and does not correspond to the true heliocentric planet.

However, the ancient Indian astronomers also introduced the notion of the *śīghrocca* for these planets whose period (see Table 1) is the same as the mean heliocentric sidereal period of these planets. Thus, in the case of the interior planets, it is the longitude of the *śīghrocca* which will be the same as the mean heliocentric longitude of the planet as understood in the currently accepted model of the solar system. As we shall see below, the traditional planetary model made use of this *śīghrocca*, crucially, in the calculation of both the longitudes and latitudes of the interior planets.

[17] W. M. Smart, *Textbook on Spherical Astronomy*, Cambridge University Press, 1965, pp. 422–23.

Name of the planet	*Āryabhaṭīya* $\frac{r_0}{R}$	*Āryabhaṭīya* Average	*Tantrasaṅgraha* $\frac{r_0}{R}$	*Tantrasaṅgraha* Average	$2e$ Modern
Sun	$\frac{13.5}{360}$	0.0375	$\frac{3}{80}$	0.0375	0.034
Moon	$\frac{31.5}{360}$	0.0875	$\frac{7}{80}$	0.0875	0.110
Mercury	$\frac{31.5-9\lvert\sin(\theta_0-\theta_m)\rvert}{360}$	0.075	$\frac{1}{6}$	0.167	0.412
Venus	$\frac{18-9\lvert\sin(\theta_0-\theta_m)\rvert}{360}$	0.0375	$\frac{1}{14+\frac{R\lvert\sin(\theta_0-\theta_m)\rvert}{240}}$	0.053	0.014
Mars	$\frac{63+18\lvert\sin(\theta_0-\theta_m)\rvert}{360}$	0.200	$\frac{7+\lvert\sin(\theta_0-\theta_m)\rvert}{39}$	0.192	0.186
Jupiter	$\frac{31.5+4.5\lvert\sin(\theta_0-\theta_m)\rvert}{360}$	0.0938	$\frac{7+\lvert\sin(\theta_0-\theta_m)\rvert}{82}$	0.091	0.096
Saturn	$\frac{40.5+18\lvert\sin(\theta_0-\theta_m)\rvert}{360}$	0.1375	$\frac{39}{360}$	0.122	0.112

Table 2: Comparison of *manda* epicyle radius and modern eccentricity.

3 *Śīghra-saṃskāra*

We shall now show that the application of *śīghra-saṃskāra* is equivalent to the transformation of the *manda-sphuṭa* to the true geocentric longitude of the planet called the *sphuṭa-graha*. Just as the *mandocca* plays a major role in the application of *manda-saṃskāra*, so too the *śīghrocca* plays a key role in the application of *śīghra-saṃskāra*. As in the case of *manda-saṃskāra*, we shall consider the application of *śīghra-saṃskāra* for the exterior and interior planets separately.

3.1 Exterior planets

For the exterior planets, Mars, Jupiter and Saturn, we have already explained that the *manda-sphuṭa-graha*, is the true heliocentric longitude of the planet. The *śīghra-saṃskāra* for them can be explained with reference to Figure 4*a*. Here A denotes the *nirayaṇa-meṣādi*, E the Earth and P the planet. The mean Sun S is referred to as the *śīghrocca* for exterior planets and thus we have

$$\begin{aligned} A\hat{S}P &= \theta_{ms} && (\textit{manda-sphuṭa}) \\ A\hat{E}S &= \theta_{s} && (\text{longitude of } \textit{śīghrocca} \text{ (mean Sun)}) \\ A\hat{E}P &= \theta && (\text{geocentric longitude of the planet}). \end{aligned}$$

The difference between the longitudes of the *śīghrocca* and the *manda-sphuṭa*, namely,

$$\sigma = \theta_s - \theta_{ms}, \tag{20}$$

is called the *śīghra-kendra* (anomaly of conjunction) in Indian astronomy. From the triangle EPS we can easily obtain the result

$$\sin(\theta - \theta_{ms}) = \frac{r_s \sin\sigma}{[(R + r_s\cos\sigma)^2 + r_s^2\sin^2\sigma]^{\frac{1}{2}}}, \tag{21a}$$

which is the *śīghra* correction formula given by Indian astronomers to calculate the geocentric longitude of an exterior planet. It may be noted that the true or geocentric longitude of the planet known as *śīghra-sphuṭa* is found in the same manner from the *manda-sphuṭa*, as the *manda-sphuṭa* is found from the mean planet, *madhyama-graha*.

From Figure 4*a* it is clear that the *śīghra-saṃskāra* transforms the true heliocentric longitudes into geocentric longitudes only if the ratio of the radii

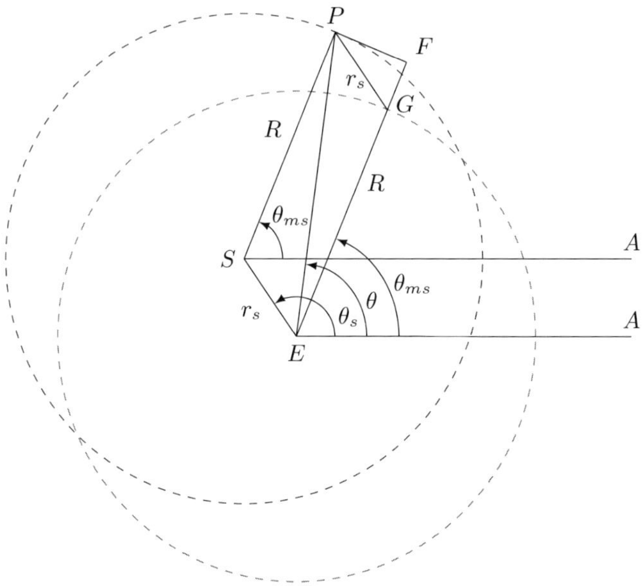

Figure 4*a*: *Śīghra* correction for exterior planets.

of the epicycle and the deferent circle is equal to the ratio of the Earth-Sun and planet-Sun distances. That this is indeed very nearly so in the Indian texts as may be seen from Table 3. But (21*a*) is still an approximation as it is based upon the mean Sun and not the true Sun.

3.2 Interior planets

The *śīghra-saṃskāra* for the interior planets can be explained with reference to Figure 4*b*. Here E is the Earth and S (*manda*-corrected mean Sun) is the *manda-sphuṭa-graha* and P, the so called *śīghrocca*, actually corresponds to the (mean heliocentric) planet. We have,

$$\begin{aligned} A\hat{E}S &= \theta_{ms} && (\textit{manda-sphuṭa}) \\ A\hat{S}P &= \theta_s && (\text{longitude of } \textit{śīghrocca}) \\ A\hat{E}P &= \theta && (\text{geocentric longitude of the planet}). \end{aligned}$$

Again, the *śīghra-kendra* is defined as the difference between the *śīghrocca* and the *manda-sphuṭa-graha* as in (20). Thus, from the triangle EPS we get the same formula

$$\sin(\theta_s - \theta_{ms}) = \frac{r_s \sin\sigma}{[(R + r_s \cos\sigma)^2 + r_s^2 \sin^2\sigma]^{\frac{1}{2}}}, \tag{21b}$$

which is the *śīghra* correction given in the earlier Indian texts to calculate the geocentric longitude of an interior planet. For the interior planets also, the value specified for $\frac{r_s}{R}$ is very nearly equal to the ratio of the planet-Sun and Earth-Sun distances as may be seen from Table 3.

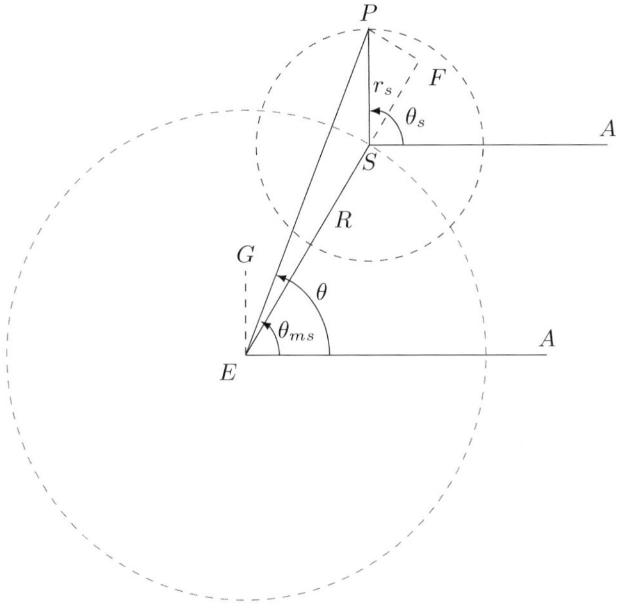

Figure 4*b*: *Śīghra* correction for interior planets.

Since the *manda* correction or equation of centre for an interior planet was applied to the longitude of the mean Sun instead of the mean heliocentric longitude of the planet, the accuracy of the computed longitudes of the interior planets according to the ancient Indian planetary models would not have been as good as that achieved for the exterior planets.

3.3 Four-step process

In obtaining the expression (21) for the *śīghra* correction, we had taken SP, the Sun-planet distance to be given by R. But, actually SP is variable and

Name of the planet	*Āryabhaṭīya* $\frac{r_s}{R}$	Average	*Tantrasaṅgraha* $\frac{r_s}{R}$	Average	*Modern* Value
Mercury	$\frac{139.5 - 9\lvert\sin(\theta_{ms} - \theta_s)\rvert}{360}$	0.375	$\frac{133 - \lvert\sin(\theta_{ms} - \theta_s)\rvert}{360}$	0.368	0.387
Venus	$\frac{265.5 - 9\lvert\sin(\theta_{ms} - \theta_s)\rvert}{360}$	0.725	$\frac{59 - 2\lvert\sin(\theta_{ms} - \theta_s)\rvert}{80}$	0.725	0.723
Mars	$\frac{238.5 - 9\lvert\sin(\theta_{ms} - \theta_s)\rvert}{360}$	0.650	$\frac{7 + \lvert\sin(\theta_{ms} - \theta_s)\rvert}{39}$	0.656	0.656
Jupiter	$\frac{72 - 4.5\lvert\sin(\theta_{ms} - \theta_s)\rvert}{360}$	0.194	$\frac{16 - \lvert\sin(\theta_{ms} - \theta_s)\rvert}{80}$	0.194	0.192
Saturn	$\frac{40.5 - 4.5\lvert\sin(\theta_{ms} - \theta_s)\rvert}{80}$	0.106	$\frac{9 - \lvert\sin(\theta_{ms} - \theta_s)\rvert}{80}$	0.106	0.105

Table 3: Comparison of $\frac{r_s}{R}$ as given in *Āryabhaṭīya* and *Tantrasaṅgraha* with the modern values of the ratio of the mean values of Earth-Sun and planet-Sun distances for the exterior planets and the inverse ratio for the interior planets.

is given by the (iterated) *manda-karṇa* K. Hence the correct form of the *śīghra* correction should be

$$\sin(\theta_s - \theta_{ms}) = \frac{r_s \sin\sigma}{\{(K + r_s\cos\sigma)^2 + r_s^2 \sin^2\sigma\}^{\frac{1}{2}}}, \tag{22}$$

where K is the (iterated) *manda-karṇa*. Since K as given by (14) and (15) depends on the *manda*-anomaly $\theta - \theta_m$, the *śīghra* correction as given by (22) cannot be tabulated as a function of the *śīghra* anomaly (σ) alone.

It has been explained in *Yuktibhāṣā* (section 8.20), that in order to simplify computation, the ancient texts on astronomy advocated that the computation of the planetary longitudes may be done using a four-step process—involving half-*manda* and half-*śīghra* corrections followed by the full-*manda* and *śīghra* corrections. The *śīghra* corrections involved in the four-step process are based on the simpler formula (21) which can be read off from a table. According to *Yuktibhāṣā*, the results of the four-step process indeed approximate those obtained by a simple sequence of *manda* correction followed by *śīghra* correction where, in the latter correction, the effect of the *manda-karṇa* is properly taken into account as in (22).

3.4 Computation of planetary latitudes

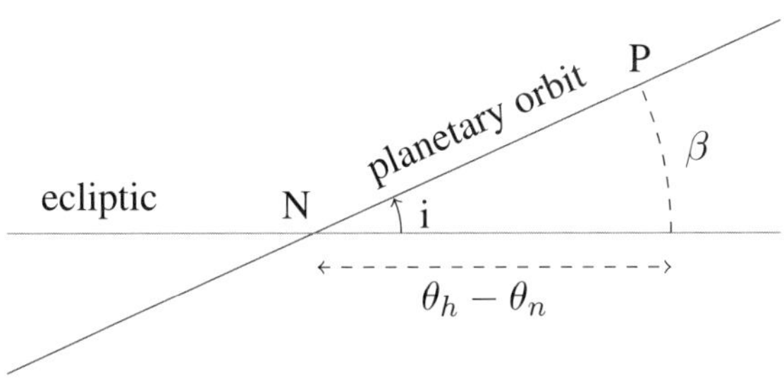

Figure 5: Heliocentric latitude of a planet.

Planetary latitudes (called *vikṣepa* in Indian astronomy) play an important role in the prediction of planetary conjunctions, occultation of stars by planets, etc. In Figure 5, P denotes the planet moving in an orbit inclined at an angle i to the ecliptic, intersecting the ecliptic at the point N, the node (called *pāta* in Indian astronomy). If β is the latitude of the planet, θ_h its heliocentric longitude, and θ_n the heliocentric longitude of the node, then it can be shown that

$$\sin\beta = \sin i \ \sin(\theta_h - \theta_n). \tag{23}$$

For small i we have

$$\beta = i \ \sin(\theta_h - \theta_n). \tag{24}$$

This is essentially the rule for calculating the latitude of a planet, as given in Indian texts, at least from the time of Āryabhaṭa.[18] For the exterior planets, it was stipulated that

$$\theta_h = \theta_{ms}, \tag{25}$$

the *manda-sphuṭa-graha* which, as we saw earlier, coincides with the heliocentric longitude of the exterior planet. The same rule applied for interior planets would not have worked, because in the traditional Indian planetary model, the *manda*-corrected mean longitude for the interior planet has nothing to do with its true heliocentric longitude. However, most of the Indian texts on astronomy stipulated that the latitude in the case of the interior planets, is to be calculated from (24) with

$$\theta_h = \theta_s + \textit{manda}\ \text{correction}, \tag{26}$$

the *manda*-corrected longitude of the *śīghrocca*. Since the longitude of the *śīghrocca* for an interior planet, as we explained above, is equal to the mean heliocentric longitude of the planet, (26) leads to the correct relation where, even for an interior planet, θ_h in (24) becomes identical with the true heliocentric longitude.

Thus, we see that the earlier Indian astronomical texts did provide a fairly accurate theory for the planetary latitudes. But they had to live with two entirely different rules for calculating latitudes, one for the exterior planets—given by (25), where the *manda-sphuṭa-graha* appears—and an entirely different one for the interior planets given by (26), which involves the *śīghrocca* of the planet, with the *manda* correction included.

This peculiarity of the rule for calculating the latitude of an interior planet was repeatedly noticed by various Indian astronomers, at least from the time of Bhāskara I (c. 629), who in his *Āryabhaṭīya-bhāṣya* drew attention to the fact that the procedure given in *Āryabhaṭīya*, for calculating the latitude of an interior planet, is indeed very different from that adopted for the exterior

[18]Equation (24) actually gives the heliocentric latitude and needs to be multiplied by the ratio of the geocentric and heliocentric distances of the planet to get the geocentric latitude. This feature was implicit in the traditional planetary models.

planets.[19] The celebrated astronomer Bhāskarācārya II (c. 1150) also draws attention to this peculiar procedure adopted for the interior planets, in his *Vāsanā-bhāṣya* on his own *Siddhāntaśiromaṇi*, and quotes the statement of *Caturveda Pṛthūdakasvāmin* (c. 860) that this peculiar procedure for the interior planets can be justified only on the ground that this is what has been found to lead to predictions that are in conformity with observations.[20]

4 Nīlakaṇṭha's revised planetary model

Amongst the available works of Nīlakaṇṭha, his revised planetary motion is discussed in the works *Tantrasaṅgraha*, *Āryabhaṭīya-bhāṣya*, *Siddhānta-darpaṇa* and the *Vyākhyā* on it, *Golasāra*, and the tract called *Grahasphuṭānayane vikṣepavāsanā*. Of these, the *Āryabhaṭīya-bhāṣya* refers to *Golasāra* and *Tantrasaṅgraha*. The *Siddhānta-darpaṇa-vyākhyā* cites the *Āryabhaṭīya-bhāṣya*. In the same way, the small but important tract *Grahasphuṭānayane vikṣepavāsanā* includes long passages from the *Āryabhaṭīya-bhāṣya* and is clearly a later composition.

The geometrical picture of planetary motion is discussed in detail in the *Āryabhaṭīya-bhāṣya*. It is also succinctly presented in terms of a few verses in both *Golasāra* and *Siddhānta-darpaṇa*. While Nīlakaṇṭha presents some aspects of his cosmological model in his *Āryabhaṭīya-bhāṣya*—in the context of discussing the geometrical picture of motion of the interior planets—a definitive but succinct account of the same is to be found in terms of a few verses in his later work *Grahasphuṭānayane vikṣepavāsanā*.

4.1 Identifying the mean of Mercury and Venus

In the very first chapter of *Tantrasaṅgraha* (c. 1500), Nīlakaṇṭha introduces a major revision of the traditional Indian planetary model, according to which what were traditionally referred to as the *śīghroccas* of the interior planets (Mercury and Venus) are now identified with the planets themselves; and the mean Sun is taken as the *śīghrocca* of all the planets.

---

[19] *Āryabhaṭīya*, with the commentary of Bhāskara I and Someśvara, K. S. Shukla (ed.) New Delhi 1976, pp. 32, 247.

[20] *Siddhāntaśiromaṇi* of Bhāskarācārya, with *Vāsanābhāṣya* and *Vāsanāvārttika* of *Nṛsiṃha Daivajña*, Muralidhara Chaturveda (ed.), Varanasi 1981, p. 402.

खाश्विदेवेषुसप्ताद्रिशराश्चेन्दोः, कुजस्य तु ।
वेदाङ्गाहिरसाङ्काश्विकराः, ज्ञस्य स्वपर्ययाः ॥
नागवेदनभस्सप्तरामाङ्कस्वरभूमयः ।
व्योमाष्टरूपवेदाङ्गपावकाश्च बृहस्पतेः ॥
अष्टाङ्गदस्रनेत्राश्विखाद्रयो भृगुपर्ययाः ।

[The number of revolutions in a *mahāyuga*] of the Moon is 5,77,53,320. That of Mars is 22,96,864. The number of own revolutions of Mercury is 1,79,37,048. That of Jupiter is 3,64,180. The number of revolutions of Venus is 70,22,268.[21]

Here the commentator Śaṅkara Vāriyar observes:

अत्र स्वशब्देन पर्यायाणां भास्कराचार्याद्यभिमतं स्वशीघ्रोच्चसम्बन्धित्वं बुधस्य निरस्तम्।

Here, by the use of the word *sva* (own), the association of this number of revolutions with the *śīghrocca* of Mercury, as done by Bhāskara and others, is rejected.[22]

It may be noted (see Table 1) that, except for the above redefinition of the mean Mercury and Venus, the *bhagaṇas* or the number of planetary revolutions in a *Mahāyuga*, are nearly the same as those given in *Āryabhaṭīya*.

4.2 Computation of planetary longitudes

Nīlakaṇṭha presents the details of his planetary model in the second chapter of *Tantrasaṅgraha*. For the exterior planets, he essentially follows the traditional model. He also retains the four-step process, while noting that (the rationale for such a scheme seems to be essentially that) such has been the recommendation of the earlier masters:

मान्दं शैघ्रं पुनर्मान्दं शैघ्रं चत्वार्यनुक्रमात् ।
कुजगुर्वर्कजानां हि कर्माण्युक्तानि सूरिभिः ॥

The earlier masters have stated that the *manda*, *śīghra* and again *manda* and *śīghra* are the four corrections that have to be applied

[21] *Tantrasaṅgraha* of Nīlakaṇṭha Somayājī, Ed. with *Laghu-vivṛti* of Śaṅkara Vāriyar by S. K. Pillai, Trivandrum 1958, p. 8.

[22] *Tantrasaṅgraha*, cited earlier, p. 9.

in sequence in the case of Mars, Jupiter and Saturn (in order to obtain their geocentric longitude).[23]

The actual procedure given by Nīlakaṇṭha is the following: If θ_0 is the mean longitude of the planet and θ_m that of its *mandocca*, then θ_1 (longitude at the end of first-step of the four-step process) is found by applying half-*manda*-correction as follows:

$$\begin{aligned}
\theta_1 &= M + \frac{1}{2} R\sin^{-1}\left[\frac{r_0}{R} R\sin(\theta_0 - \theta_m)\right], \qquad \text{with} \\
\frac{r_0}{R} &= \frac{[7 + |\sin(\theta_0 - \theta_m)|]}{39} \qquad \text{(for Mars)} \\
\frac{r_0}{R} &= \frac{[7 + |\sin(\theta_0 - \theta_m)|]}{82} \qquad \text{(for Jupiter)} \\
\frac{r_0}{R} &= \frac{39}{320} \qquad \text{(for Saturn).}
\end{aligned}$$

Then, θ_2 is found by applying half-*śīghra* -correction with the mean Sun θ_s as the *śīghrocca* as follows:

$$\begin{aligned}
\theta_2 &= \theta_1 + \frac{1}{2} R\sin^{-1}\left[\frac{r_s}{K_{s1}} R\sin(\theta_1 - \theta_s)\right], \qquad \text{with} \\
K_{s1} &= [\{r_s \sin(\theta_1 - \theta_s)\}^2 + \{R + r_s \cos(\theta_1 - \theta_s)\}^2]^{\frac{1}{2}} \\
\left(\frac{r_s}{R}\right) &= \frac{[53 - 2|\sin(\theta_1 - \theta_s)|]}{80} \qquad \text{(for Mars)} \\
\left(\frac{r_s}{R}\right) &= \frac{[16 - |\sin(\theta_1 - \theta_s)|]}{80} \qquad \text{(for Jupiter)} \\
\left(\frac{r_s}{R}\right) &= \frac{[9 - |\sin(\theta_1 - \theta_s)|]}{80} \qquad \text{(for Saturn).}
\end{aligned}$$

Then the *manda-sphuṭa* θ_{ms} is found by adding the whole *manda*-correction obtained with θ_2 to θ_0

$$R\sin(\theta_{ms} - \theta_0) = \left(\frac{r_0}{R}\right) R\sin(\theta_2 - \theta_m).$$

Then the true planet *sphuṭa-graha* P is found by applying the whole of *śīghra*-correction to θ_{ms}.

$$R\sin(\theta - \theta_{ms}) = \left[\frac{r_s}{K_s} R\sin(\theta_{ms} - \theta_s)\right]$$

[23] *Tantrasaṅgraha*, cited earlier, p. 41.

$$\text{where} \quad K_s = [\{r_s \sin(\theta_{ms} - \theta_s)\}^2 + \{R + r_s \cos(\theta_{ms} - \theta_s)\}^2]^{\frac{1}{2}}. \tag{27}$$

Again, as we had noted earlier in connection with the traditional planetary model, in the above four-step process also, the iterated *manda*-hypotenuse (*aviśiṣṭa-manda-karṇa*) does not appear and the *manda* and *śīghra*-corrections can be read off from a table.

In the case of interior planets, Nīlakaṇṭha presents just the two-step process: *manda-saṃskāra* followed by the *śīghra-saṃskāra*. For the interior planets, if θ_0 is the mean planet (as per his revised model), U its *mandocca* and S is the mean Sun (*śīghrocca*), then the *manda*-correction leading to the *mandasphuṭa*, is given by

$$\begin{aligned} R\sin(\theta_{ms} - \theta_0) &= \frac{r_0}{R} R\sin(\theta_0 - \theta_m) \\ \frac{r_0}{R} &= \frac{1}{6}, \frac{1}{\left[14 + \frac{|R\sin(\theta_0 - \theta_m)|}{240}\right]} \qquad \text{(for Mercury, Venus).} \end{aligned}$$

It may be recalled that the *aviśiṣṭa-manda-karṇa* K is to be calculated using the Mādhava formula (15). The *śīghra*-correction giving the true planet θ is given by,

$$\begin{aligned} R\sin(\theta - \theta_s) &= \left[\left(\frac{r_s}{R}\right)\left(\frac{K}{K_s}\right) R\sin(\theta_{ms} - \theta_s)\right] \\ \text{where} \quad K_s &= [R\sin(\theta_{ms} - \theta_s)^2 + \{R\cos(\theta_{ms} - \theta_s) + \left(\frac{r_s}{R}\right) K\}^2]^{\frac{1}{2}} \\ \left(\frac{r_s}{R}\right) &= \frac{[31 - 2|\sin(\theta_{ms} - \theta_s)|]}{80R} \qquad \text{(for Mercury)} \\ \left(\frac{r_s}{R}\right) &= \frac{[59 - 2|\sin(\theta_{ms} - \theta_s)|]}{80R} \qquad \text{(for Venus).} \end{aligned} \tag{28}$$

Note that in the above two-step process, the *aviśiṣṭa-manda-karṇa* K shows up in the *śīghra*-correction. In his discussion of the geometrical picture of planetary motion in the *Āryabhaṭīya-bhāṣya*, Nīlakaṇṭha presents the two-step process as the planetary model for all planets. This has also been the approach of *Yuktibhāṣā*.

4.3 Planetary latitudes

In the seventh chapter of *Tantrasaṅgraha*, Nīlakaṇṭha gives the method for calculating the latitudes of planets, and prescribes that for all planets, both exterior and interior, the latitude is to be computed from the *manda-sphuṭa-graha*.

> मन्दस्फुटात् स्वपातोनात् भौमादीनां भुजागुणात्।
> परमक्षेपनिघ्ना स्यात् क्षेपोऽन्त्यश्रवणोद्धृतः॥
>
> The Rsine of the *manda-sphuṭa* of the planet Mars etc., from which the longitude of its node is subtracted, is multiplied by the maximum latitude and divided by the last hypotenuse (the *śīghra* hypotenuse of the last step). The result is the latitude of the planet.[24]

This is as it should be for, in Nīlakaṇṭha's model, the *manda-sphuṭa-graha* (the *manda*-corrected mean longitude) coincides with the true heliocentric longitude for both the exterior and interior planets. In this way, Nīlakaṇṭha, by his modification of traditional Indian planetary theory, solved the long-standing problem in Indian astronomy, of there being two different rules for calculating the planetary latitudes.

It is noteworthy that in the above verse, Nīlakaṇṭha states that the last hypotenuse that arises in the process of computation of longitudes, namely the *śīghra-karṇa* K_s is to be used as the divisor. In *Āryabhaṭīya-bhāṣya*, he identifies this as the Earth-planet distance (*bhū-tārāgraha-vivara*). There, Nīlakaṇṭha has also explained how the computations of true longitude and latitude get modified when latitudinal effects are also taken into account. The true Earth-planet distance (*bhū-tārāgraha-vivara*) is also calculated there in terms of the K_s and the latitude.[25]

4.4 Rationale for the revised planetary model

In his *Āryabhaṭīya-bhāṣya*, Nīlakaṇṭha explains the rationale behind his revision of the traditional planetary theory. This has to do with the fact (which,

[24] *Tantrasaṅgraha*, cited earlier, p. 139.

[25] *Āryabhaṭīya-bhāṣya* of Nīlakaṇṭha ed. by K. Sāmbaśiva Sāstrī, Vol. III, Trivandrum, 1957, pp. 6-7. This issue has also been discussed at great length in *Gaṇita-yukti-bhāṣā* (cited earlier, Vol. II, Section 8.16, pp. 495–500, 653–59, 883–89).

as we have mentioned above, was also noticed by several Indian astronomers prior to Nīlakaṇṭha) that the traditional planetary model employed entirely different schemes for computing the latitudes of the exterior and the interior planets. While the latitude of the exterior planets was computed from their so called *manda-sphuṭa* (which corresponds to what we currently refer to as the true heliocentric planet), the latitudes of the interior planets was computed from their so-called *śīghrocca*. Nīlakaṇṭha argued that since the latitude should be dependent upon the deflection (from the ecliptic) of the planet itself and not of any other body, what was traditionally referred to as the *śīghrocca* of an interior planet should be identified with the planet itself. Nīlakaṇṭha also showed that this would lead to a unified treatment of the latitudinal motion of all the planets—interior as well as the exterior.

In his commentary on the verse 3 of *Golapāda* of Āryabhaṭa dealing with the calculation of latitudes, Nīlakaṇṭha discusses the special features that arise in the case of interior planets. It is here that he provides a detailed rationale for his revision of the traditional planetary model:

> शीघ्रवशाच्च विक्षेप उक्तः। कथमेतद्युज्यते? ननु स्वबिम्बस्य विक्षेपः स्वभ्रमणवशादेव भवितुमर्हति, न पुनः अन्यभ्रमणवशादिति। सत्यम्। न पुनः अन्यस्य भ्रमणवशात् अन्यस्य विक्षेपः उपपद्यते। तस्मात् बुधः अष्टाशीत्यैव दिनैः स्वभ्रमणवृत्तं पूरयति।...
>
> एतच्च नोपपद्यते यदेकेनैव संवत्सरेण तत्परिभ्रमणमुपलभ्यते नैवाष्टाशीत्या दिनैः। सत्यम्। भगोलपरिभ्रमणं तस्याप्येकेनैवाब्देन।...
>
> एतदुक्तं भवति। तयोः भ्रमणवृत्तेन न भूः कबलीक्रियते। ततो बहिरेव सदा भूः। भगोलैकपार्श्वे एव तद्वृत्तस्य परिसमाप्तत्वात् तद्भगणेन न द्वादशराशिषु चारः स्यात्। तयोरपि वस्तुतः आदित्यमध्यम एव शीघ्रोच्चम्। शीघ्रोच्चभगणत्वेन पठिता एव स्वभगणाः। तथापि आदित्यभ्रमणवशादेव द्वादशराशिषु चारः स्यात्, शीघ्रवृत्तस्य कक्ष्यायाः महत्त्वात्। शीघ्रोच्चनीचवृत्तस्याप्येकभागगमेव स्वभ्रमणवृत्तम्। यथा कुजादीनामपि शीघ्रोच्चं

स्वमन्दकक्ष्यामण्डलादिकमाकर्षति एवमेतयोरपि।
अनयोः पुनः तदाकर्षणवशादेव द्वादशराशिषु चार इति।

The latitudinal motion is said to be due to that of the *śīghrocca*. How is this appropriate? Isn't the latitudinal motion of a body dependent on the motion of that body only, and not because of the motion of something else? The latitudinal motion of one body cannot be obtained as being due to the motion of another. Hence [we should conclude that] Mercury goes around its own orbit in 88 days ... However this also is not appropriate because we see it going around [the Earth] in one year and not in 88 days. True, the period in which Mercury completes one full revolution around the *bhagola* (the celestial sphere) is one year only [like the Sun]. ...

All this can be explained thus: Their [Mercury and Venus] orbits do not circumscribe the Earth. The Earth is always outside their orbit. Since their orbit is always confined to one side of the [geocentric] celestial sphere, in completing one revolution they do not go around the twelve signs (*rāśis*). For them also really the mean Sun is the *śīghrocca*. It is only their own revolutions, which are stated to be the revolutions of the *śīghrocca* [in the *Āryabhaṭīya*]. It is only due to the revolution of the Sun [around the Earth] that they (i.e. the interior planets, Mercury and Venus) complete their movement around the twelve signs [and complete their revolution of the Earth]. Because the *śīghra*-epicycle is larger than their orbit. Their orbit is completed on one side of the *śīghra*-epicycle. Just as in the case of Jupiter etc. [exterior planets] the *śīghrocca* attracts [and drags around] the *manda*-orbits on which they move (*manda-kakṣyā-maṇḍala*), in the same way it does for these [interior] planets also. And it is due to this attraction that these [interior planets] move around the twelve signs.[26]

[26] *Āryabhaṭīya-bhāṣya* of Nīlakaṇṭha, Ed by, Vol. III, p. 8-9.

5 Geometrical picture of planetary motion according to Nīlakaṇṭha

5.1 Exterior planets

Nīlakaṇṭha first gives the following general outline of the geometrical picture of planetary motion:

> अत्रायमभिसन्धिः। कक्ष्यामण्डलकेन्द्र एव शीघ्रपरिधेरपि केन्द्रम्। तत्परिधौ शीघ्रोच्चाक्रान्तप्रदेशे मन्दपरिधिकेन्द्रं च। एवं परिधौ पुनर्मन्दोच्चप्रदेशे प्रतिमण्डलकेन्द्रं च। तच्च प्रतिमण्डलमाकाशकक्ष्यायाः स्वभगणावाप्तैर्योजनैस्तुल्यम्। तस्मिन्नेव ग्रहबिम्बं इतरैः समयोजनगतिर्भ्रमति। तत्तुल्यमेव तत् कक्ष्यामण्डलं शीघ्रपरिधौ उच्चप्रदेशे केन्द्रं कृत्वा परिलेखनीयम्। तत्रापि कर्णमण्डलं मन्दकर्णन्यायेन अविशेष्य परिलेखनीयम्।
>
> Here, what is intended to be conveyed is as follows: The centre of the *kakṣyā-maṇḍala* (concentric) is also the centre of the *śīghra* epicycle; on that epicycle, at the location of *śīghrocca*, is the centre of the *manda* epicycle; in the same way, on that *manda* epicycle at the location of mandocca is the centre of the *pratimaṇḍala* (eccentric). (The circumference of) that *pratimaṇḍala* is equal to the circumference of the sky (*ākāśa-kakṣyā*) divided by the revolution number of the planet. It is only in that (*pratimaṇḍala*) that the planetary orb moves with a constant linear velocity. The corresponding concentric (*kakṣyā-maṇḍala*) should be drawn with the same dimension with its centre on the *śīghra* epicycle at the location of *śīghrocca*. There also the circle of the hypotenuse is to be obtained by the process of iteration as per the rule for the *manda-karṇa*.[27]

Later, while commenting on the verse 3 of *Golapāda*, Nīlakaṇṭha explains how the above picture needs to be modified when the latitudinal motion is also taken into account. The main feature is that it is the *manda* epicycle together with the eccentric which is inclined to the ecliptic and not the *śīghra* epicycle (which represents the Earth-Sun relative motion):

[27] *Āryabhaṭīya-bhāṣya* of Nīlakaṇṭha cited above, Vol. II, p. 70.

भगोलमध्यनाभिकस्य कार्त्स्न्येन अपमण्डलमार्गगस्य शीघ्रवृत्तस्य परिधौ यः शीघ्रोच्चसमप्रदेशः तद्धि मन्दकर्मणि कक्ष्यामण्डलकेन्द्रमिति कालक्रियापादे एवे उक्तम्। तदेव मन्दोच्चनीचवृत्तस्य कर्णमण्डलस्य च केन्द्रम्। एवमेतनि त्रीणि मण्डलानि अपमण्डलमार्गमभितः अर्धशः उत्तरतो दक्षिणतश्च विक्षिप्तानि।

> It has already been stated in the *Kālakriyāpāda* that on the *śīghra-vṛtta*, which has its centre at the centre of the celestial sphere and is in the plane of the ecliptic, the point which corresponds to the *śīghrocca* is in fact the centre of the *kakṣyā-maṇḍala* (concentric) in the *manda*-process. The same (*śīghrocca*) is also the centre of the *manda-nicocca-vṛtta* (the *manda*-epicycle) and also of the (*manda*) *karṇa-maṇḍala* (hypotenuse circle or the orbit). In this way these three circles (*manda*-concentric, epicycle and the hypotenuse circle) are inclined to the ecliptic both towards the north and the south.[28]

Based on the description presented above, we arrive at the geometrical picture of motion—for an exterior planet—as shown in Figure 6*a*. In this figure, O represents the location of the observer and is considered to be *bhagola-madhya* (centre of the celestial sphere). The circle centered around O, with radius equal to tabulated radius of *śīghra* epicycle, r_s, is called *śīghra-nīcocca-vṛtta*, on which the *śīghrocca* or the mean Sun S is located.

It is said that the *manda-nīcocca-vṛtta* (also called *mandaparidhi*) is a circle with the *śīghrocca* as the centre. The *mandocca* U is located on this circle, whose radius is equal to the (variable) radius of the *manda* epicycle. The *pratimaṇḍala* on which the planet P moves is centered at the *mandocca*. SP is the *manda-karṇa* denoted by K and $\Gamma\hat{S}P$ is the *manda-sphuṭa*. $\Gamma\hat{O}P$ is the true geocentric planet known as the *śīghra-sphuṭa*. The distance of the planet from the centre of the *bhagola* is denoted by K_s and it is also the *śīghra-karṇa*.

Among the various circles depicted in Figure 6*a*, it is said, that the circles centered around the *śīghrocca* S, namely the *manda-nīcocca-vṛtta*, the *manda-karṇa-vṛtta* and the *manda* concentric (which is not indicated in the figure), are inclined to the plane of the ecliptic towards the north and the south. The figure also depicts a section of the *śīghra-karṇa-vṛtta*—centered

[28] *Āryabhaṭīya-bhāṣya* of Nīlakaṇṭha cited above, Vol III, p. 5.

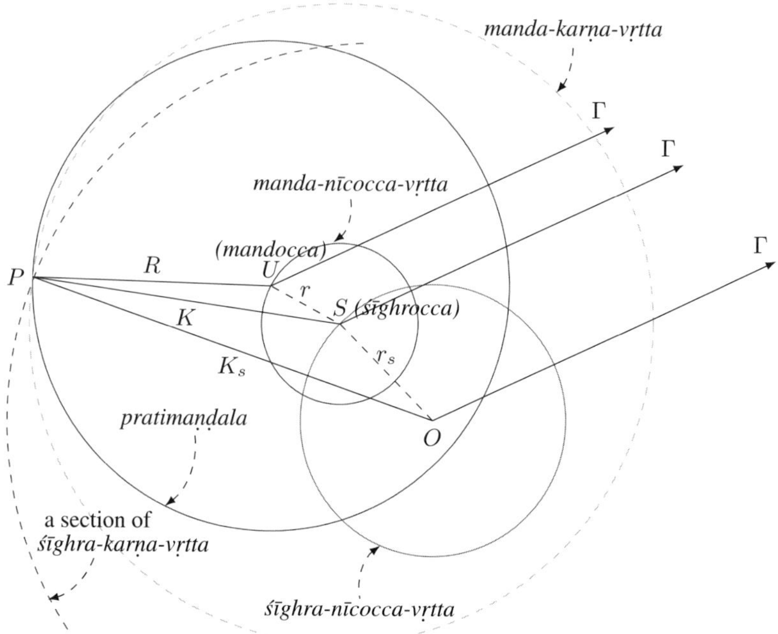

Figure 6*a*: Geometrical picture of motion of an exterior planet given by Nīlakaṇṭha.

around *O*—which represents the instantaneous orbit (the orbit in which the planet moves at that instant) of the planet with respect to the Earth.

5.2 Interior planets

Nīlakaṇṭha explains in the commentary on verse 3 of *Golapāda* that the above geometrical picture of motion needs to be modified in the case of interior planets. We have earlier (in section 4.4) cited a part of this discussion where Nīlakaṇṭha had noted that the interior planets go around the Sun in orbits that do not circumscribe the Earth, in a period that corresponds to the period of their latitudinal motion, and that they go around the zodiac in one year as they are dragged around the Earth by the Sun. Having identified the special feature of the orbits of the interior planets that they do not circumscribe the Earth, Nīlakaṇṭha explains that it is their own orbit, which is smaller than the *śīghra-nīcocca-vṛtta*, that is tabulated as the epicycle in a measure where the latter is 360 degrees.

तेनोभे अपि वृत्ते व्यत्यस्य कल्प्येते। तयोः शीघ्रोच्चनीचवृत्तस्य षष्टिशतत्रयांशेनैव परिमाय स्वकक्ष्यावृत्तमेव शीघ्रोच्चनीचवृत्तत्वेन कल्प्यते। मन्दोच्चनीचवृत्तञ्च तदंशेनैव परिमाय पठितम्।

These two circles (concentric and the *śīghra* epicycle) are now to be imagined in the contrary way. Of them, the concentric itself (being smaller than the epicycle) is given in units where the *śīghra* epicycle is taken to be three hundred and sixty degrees, and will now play the role of epicycle. The *manda* epicycle is also taken to be tabulated in terms of this (concentric).[29]

Nīlakaṇṭha then goes on to explain the process of computation of the true longitude of these planets in the same manner as outlined in *Tantrasaṅgraha* and corresponding to the following geometrical picture of motion.

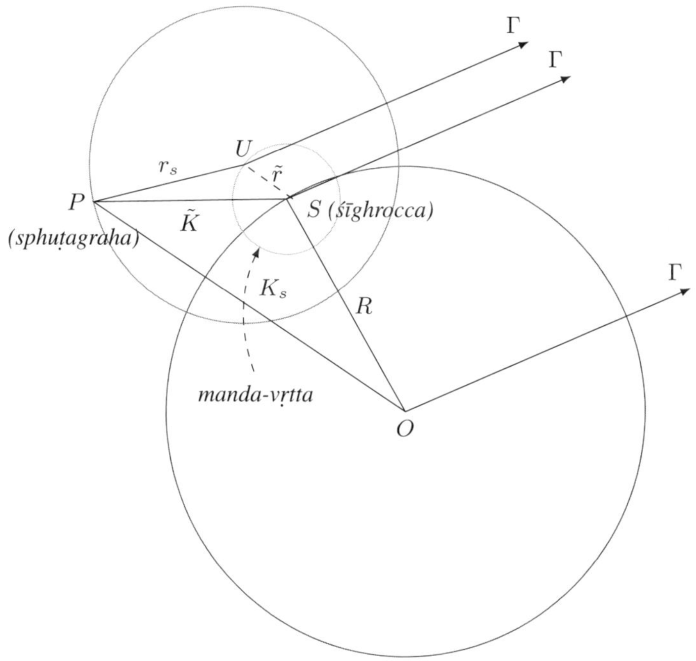

Figure 6*b*: Geometrical picture of motion of an interior planet given by Nīlakaṇṭha.

[29] *Āryabhaṭīya-bhāṣya* of Nīlakaṇṭha cited above, Vol III, p. 9.

The geometrical picture of motion of the interior planets as presented by Nīlakaṇṭha is shown in Figure 6*b*. Here, O is the observer, assumed to be at the centre of the celestial sphere (*bhagola-madhya*). S is the *śīghrocca* which is taken to be the mean Sun for all the planets. P is the planet moving around the mean Sun in an eccentric orbit. This eccentric orbit is centered at U, the *mandocca*. The point U itself is conceived to be moving on the *manda-nīcocca-vṛtta* centered around S.

For interior planets the planet-Sun distance is smaller than the Earth-Sun distance. Hence, the radius of the planet's eccentric orbit (UP) is taken to be the radius of the *śīghrocca-nīca-vṛtta* r_s, and the radius of the mean Sun's orbit (OS) is taken to be *trijyā* R. Further, since the *mandapratimaṇḍala*, or the *manda* eccentric on which the planet moves is of radius r_s and not R, the radius (variable) of the *manda* epicycle r itself is to be scaled by a factor $\frac{r_s}{R}$ and will be $\tilde{r} = r\frac{r_s}{R}$. Correspondingly the (iterated) *manda-karṇa* K will also be scaled to $\tilde{K} = K\frac{r_s}{R}$.

6 Nīlakaṇṭha's cosmological model

While discussing the geometrical picture of planetary motion in his *Āryabhaṭīya-bhāṣya* as well as *Golasāra* and *Siddhānta-darpaṇa*, Nīlakaṇṭha considers the orbits of each of the planets individually and does not put them together in a single cosmological model of the planetary system. However, he presents his cosmological model very clearly in a remarkable short tract called *Grahasphuṭānayane vikṣepavāsanā*, which seems to have been written after the *Āryabhaṭīya-bhāṣya* as the former quotes extensively from the latter. Here he clearly integrates the geometrical picture of motion of different planets into a single model of the planetary system by identifying the *śīghrocca*, around which each of the planets rotates, with the physical "mean Sun moving on the orbit of the Sun". Based on this identification, Nīlakaṇṭha also states that the ratio of the radius of the *śīghra* epicycle to that of the concentric is nothing but the ratio of the mean radius of the orbit of the Sun around the Earth to the mean radius of the orbit of the planet itself, in the case of the exterior planets; and the other way around, in the case of the interior planets. He further explains that this difference between the exterior and interior planets is caused, in the case of the interior planets, by their orbit being smaller than the orbit of the Sun around the Earth and the dimensions of the epicycle and concentric having to be interchanged. In Nīlakaṇṭha 's own words:

इन्द्वादेः स्वस्वपातद्वयत उदगवाग् अर्धशः क्रान्तिवृत्तात्
विक्षिप्ता मान्दकक्ष्या कथितनिजलवैः सर्वदा तुल्यसङ्ख्यैः।
तत्रेन्दोर्मान्दकक्ष्या ह्यपमवलयमध्यस्थकेन्द्रा कुजादे-
र्मान्दाः कक्ष्या भगोलस्थितदिनकरकक्ष्यास्थमध्यार्ककेन्द्राः॥

किञ्चारेड्यार्कजानां निजवृतिकलया मापयित्वार्ककक्ष्यां
शैघ्राण्युक्तानि वृत्तानि हि बुधसितयोस्त्वर्ककक्ष्याकलाभिः।
स्वां कक्ष्यां मापयित्वा पुनरिह कथिते शीघ्रवृत्ते...

The *manda-vṛttas* of the Moon and the others (the five planets) are deflected from the two nodes of their own orbits, half way towards the north and the south of the ecliptic (*krānti-vṛtta*) by a measure that has been specified separately [for each planet] and which remains the same for all times. There [again] the *manda-vṛtta* of the Moon is centered at the center of the ecliptic (*apamavalaya*), where as the *manda-vṛttas* of Mars, etc. (the five planets), are centered at the mean Sun which lies on the orbit of the Sun (*dinkara-kakṣyāstha-madhyārka*) situated in the celestial sphere (*bhagola*).

Moreover, in the case of Mars, Jupiter and Saturn, the [dimensions of their] *śīghra-vṛttas* have been stated by measuring the orbit of the [mean] Sun (*arka-kakṣyā*) in terms of minutes of (the dimensions of) their own orbits (*nija-vṛti-kalayā*). However in the case of Mercury and Venus, the [dimensions of their] *śīghra-vṛttas* have indeed (*punaḥ*) been stated by measuring their own orbits in terms of the minutes of (the dimension of) the orbit of the [mean] Sun (*arka-kakṣyā-kalābhiḥ*). ...[30]

The first verse clearly describes the cosmological model of Nīlakaṇṭha, which is that the five planets, Mercury, Venus, Mars, Jupiter and Saturn, go around the mean Sun in an eccentric orbit—inclined to the ecliptic (see Figure 7)—while the mean Sun itself goes around the Earth.[31] It is in the

[30]*Grahasphuṭānayane vikṣepavāsanā* in *Gaṇitayuktayaḥ* Ed. K. V. Sarma, Hoshiarpur 1979, p. 58. As we noted earlier, the initial verses of the anonymous tract *Vikṣepavāsanā*, closely follow the above verses of Nīlakaṇṭha.

[31]As we noted earlier, this cosmological model is the same as the one proposed by Tycho Brahe, albeit on entirely different considerations, towards the end of sixteenth century.

second verse that Nīlakaṇṭha makes the remarkable identification that

$$\frac{r_s}{R} = \frac{\text{mean Earth-Sun distance}}{\text{mean Sun-planet distance}} \quad \text{(for exterior planets)} \qquad (29a)$$

$$\frac{r_s}{R} = \frac{\text{mean Sun-planet distance}}{\text{mean Earth-Sun distance}} \quad \text{(for interior planets).} \qquad (29b)$$

where r_s is the radius of the *śīghra* epicycle and R is the radius of the concentric. We had noted in Chapter II that the *śīghra*-process serves to transform the heliocentric longitudes to geocentric longitudes, precisely because the above relations (29*a* and 29*b*) are indeed satisfied (see Table 3), even though the traditional Indian astronomical texts did not conceive of any such relation between the radii of the *śīghra* epicycles and the mean ratios of Earth-Sun and Sun-planet distances. In fact, Nīlakaṇṭha seems to be the first Indian astronomer to explicitly state the relations (29*a* and 29*b*), which seems to follow clearly from his identification of the *śīghrocca* of each planet with the physical "mean Sun lying on the orbit of the Sun" (*dinakara-kakṣyāstha-madhyārka*).[32]

7 Concluding Remarks

From the above discussion it is clear that the central feature of Nīlakaṇṭha's revision of the traditional planetary model is that the *manda*-correction or the equation of centre for the interior planets, should be applied to the mean heliocentric planet (or what was referred to as *śīghrocca* in the traditional Indian planetary model), and not the mean Sun. In this way Nīlakaṇṭha, by 1500 CE, had arrived at the correct formulation of the equation of centre for the interior planets, perhaps for the first time in the history of astronomy. Nīlakaṇṭha was also able to formulate a unified theory of planetary latitudes.

Just as was the case with the earlier Indian planetary model, the ancient Greek planetary model of Ptolemy and the planetary models developed in the Islamic tradition during the 8th-15th centuries postulated that the equation of centre for an interior planet should be applied to the mean Sun, rather than to the mean heliocentric longitude of the planet as we understand today. In fact, Ptolemy seems to have compounded the confusion by clubbing together

[32] As we noted earlier, Nicholas Copernicus also seems to have arrived at the same relation (around the same time as Nīlakaṇṭha) by identifying the epicycle associated with so called "solar anomaly" in the Ptolemaic model with the orbit of the Earth around the Sun in the case of the exterior planets and with the orbit of the planet itself in the case of the interior planets.

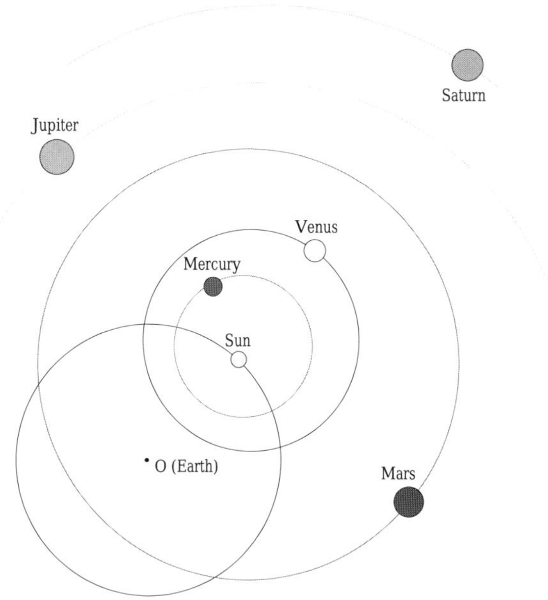

Figure 7: Nīlakaṇṭha's cosmological model showing the five planets moving in eccentric orbits around the mean Sun.

Venus along with the exterior planets and singling out Mercury as following a slightly deviant geometrical model of motion.[33] Further, while the ancient Indian astronomers successfully used the notion of the *śīghrocca* to arrive at a satisfactory theory of the latitudes of the interior planets, the Ptolemaic model is totally off the mark when it comes to the question of latitudes of these planets.[34]

Even the celebrated Copernican revolution brought about no improvement

[33]See for example, *The Almagest by Ptolemy*, Translated by G. J. Toomer, London 1984. For the exterior planets, the ancient Indian planetary model and the model described by Ptolemy are very similar except that, while the Indian astronomers use a variable radius epicycle, Ptolemy introduces the notion of an equant. Ptolemy adopts the same model for Venus also, and presents a slightly different model for Mercury. In both cases the equation of centre is applied to the mean Sun.

[34]As a well known historian of astronomy has remarked: "In no other part of planetary theory did the fundamental error of the Ptolemaic system cause so much difficulty as in accounting for the latitudes, and these remained the chief stumbling block up to the time of Kepler" (J. L. E. Dreyer, A History of Astronomy from Thales to Kepler, New York 1953, p. 200).

in the planetary theory for the interior planets. As is widely known now,[35] the Copernican model was only a reformulation of the Ptolemaic model (with some modifications borrowed from the Maragha School of Astronomy of Nasir ad-Din at-Tusi (c. 1201-74), Ibn ash-Shatir (c. 1304-75) and others) for a heliocentric frame of reference, without altering his computational scheme in any substantial way for the interior planets. Thus, it appears that the correct rule for applying the equation of centre for an interior planet to the mean heliocentric planet (as opposed to the mean Sun), and a satisfactory theory of latitudes for the interior planets, were first formulated in the Greco-European astronomical tradition only in the early 17th century by Kepler.

However, it was seen above that the traditional Indian planetary model presented a fairly accurate computational scheme for calculating the longitudes and latitudes for the exterior planets. With his revision of the traditional model, Nīlakaṇṭha arrived at a fairly accurate scheme for the interior planets also. In fact, as far as the computational scheme for calculating the planetary longitudes and latitudes is concerned, Nīlakaṇṭha 's model is indeed a good approximation to the Keplerian model of planetary motion.

[35] As an important study (N. M Swerdlow and O. Neugebauer, Mathematical Astronomy in Copernicus' De Revolutionibus, Part I, New York 1984, p. 483) notes: "Copernicus, ignorant of his own riches, took it upon himself for the most part to represent Ptolemy, not nature, to which he had nevertheless come the closest of all". In this famous and just assessment of Copernicus, Kepler was referring to the latitude theory of Book V [of De Revolutionibus], specifically to the "librations" of the inclinations of the planes of the eccentrics, not in accordance with the motion of the planet, but by the unrelated motion of the Earth. This improbable connection between the inclinations of the orbital planes and the motion of the Earth was the result of Copernicus's attempt to duplicate the apparent latitudes of Ptolemy's models in which the inclinations of the epicycle planes were variable. In a way this is nothing new since Copernicus was also forced to make the equation of centre of the interior planets depend upon the motion of the Earth rather than the planet."

Rythmes et algorithmes
Le génie mathématique indien

François Patte
Laboratoire MAP5 (UMR CNRS 8145), Université Paris Descartes

En expliquant une règle de calcul combinatoire donnée par Bhāskara II, le commentateur Gaṅgādhara présente en exemple une suite de nombres, et un moyen mémotechnique, versifié, pour se souvenir des premiers nombres de la suite et du processus de construction de celle-ci.

Voici cette suite :

1, 2, 3, 6, 10, 19, 33, 60, 106, 191, 340, 610, ...

Et voici le poème mnémotechnique :

> Les combinaisons jusqu'à six demi-brèves se mémorisent par : un *(lakṣmī)*, deux, trois (feu), six (membres), dix (directions), dix-neuf ; les deux derniers additionnés et aussi les deux [séparés] par un intervalle, on aura aussi le dénombrement avec la somme des uns et des autres[1].

Une suite de nombres parle rarement d'elle-même ; quant au poème, il est plutôt obscur et, sans clef, ne permet pas de comprendre cette suite.

La solution de l'énigme se trouve dans un traité de musique du XIVe siècle dû à Śārṅgadeva : le *Saṃgītaratnākara* (l'océan de la musique).

Le chapitre cinq de cette œuvre est consacré à l'étude du rythme ; on y trouve le moyen d'énumérer exhaustivement, et de manière ordonnée, tous

[1] *Lakṣmīdvivahnyaṅgadigekakonaviṃśatyabhijñā drutaṣaṭkabandhaḥ* | *antyau yutau dvāv atha sāntarau dvau tattadyuteḥ syāt purato 'pi saṃkhyā* ‖ Ce poème utilise le système des *bhūtasaṃkhyā* (nombres qui sont des choses) pour désigner certains nombres : *lakṣmī* semble désigner le nombre « un », *vahni*, feu, est le nombre « trois », car il faut trois feux dans le sacrifice à Agni, le Veda a six auxiliaires (membres, *aṅga*), il y a dix directions dans l'espace *(diś)*.

les modèles rythmiques d'une durée déterminée utilisant quatre figures de notes. Une fois l'énumération faite, on donne le moyen de compter le nombre de variantes rythmiques obtenues et celui de retrouver un modèle connaissant sa place dans l'énumération ou, inversement, de calculer le rang d'un modèle donné.

Le *Saṃgītaratnākara* utilise quatre figures de notes dont les rapports de durée sont : 1, 2, 4 et 6 :

- *druta*, demi-brève[2], noté : **o**[3]. C'est la plus petite durée, elle sert aussi d'unité pour donner la longueur des mesures musicales : on dira une mesure de quatre, cinq, sept *druta*. Dans le rapport des durées utilisées ici, on peut la représenter par une croche ♪
- *laghu*, brève[4], notée : **I**. D'une durée de deux *druta*, elle correspond à la noire ♩
- *guru*, longue[5], notée : **S**. Elle vaut deux *laghu*, soit quatre *druta* et correspond à une blanche 𝅗𝅥
- *pluta*, protractée, notée **S̀**, elle a une durée de trois *laghu*, c'est-à-dire six *druta*, et correspond donc à une blanche pointée 𝅗𝅥.

Prastāra

Prastāra signifie « étendu ». C'est par ce nom que les traités de prosodie et de musique désignent le tableau dont chaque ligne représente les variations rythmiques qui peuvent être construites pour une durée donnée. Il désigne aussi l'algorithme qui permet de construire ce tableau.

Dans le cas des rythmes musicaux, un *prastāra* pour quatre *druta* présentera toutes les variations rythmiques d'une durée de quatre demi-brèves, par exemple : **o o o o** (♪♪♪♪), **I I** (♩ ♩) ou **o I o** (♪♩ ♪), etc.

Pour faire cette construction de manière exhaustive, Śārṅgadeva propose l'algorithme suivant :

> *Après avoir posé une plus petite au-dessous de la première plus grande, le reste est comme au-dessus. S'il y a un manque au début, on écrira, si cela est possible, de plus grandes sur la gauche — en cas d'impossibilité des petites —, pour compléter le rythme ; cette prescription, dont le terme est [un rythme*

[2]*litt.* rapide

[3]Dans cet article, les notations employées pour les figures de notes imitent les notations utilisées dans les manuscrits sanskrits.

[4]*litt.* légère.

[5]*litt.* lourde.

composé] entièrement de demi-brèves, doit être exécutée à plusieurs reprises. Cette permutation existe quand il y a brèves, longues ou protractées, isolées ou composées, mais pas quand il y a une demi-brève isolée[6].

Le *prastāra* se construira de haut en bas et la règle explique comment construire une ligne au-dessous d'une autre.

« Plus petite... », « plus grande... », le texte paraît peu clair. C'est un mode d'expression, largement développé par les mathématiciens indiens, qui permet la construction de règles itératives pour résoudre des problèmes ou, comme c'est le cas ici, pour décrire un algorithme. Le recours aux commentaires est presque indispensable et celui de Kallinātha *(Kalānidhi)* permet de bien comprendre la procédure décrite ; voici son explication pour la première phrase :

> « Après avoir écrit au début un quelconque rythme de son choix, [on posera au-dessous] de la première plus grande parmi ses composants... Ici, par le mot « plus grand » on entend brève, longue ou protractée. En voici le sens : quand la première plus grande est une brève alors une demi-brève, qui est plus petite par rapport à celle-là, doit être écrite au-dessous. Mais quand la première plus grande est une longue, alors une brève, qui est plus petite par rapport à celle-là, doit être écrite au-dessous. Quand, encore, la première plus grande est une protractée, une longue, qui est plus petite par rapport à celle-là, doit être écrite au-dessous. Ainsi, ce qui est invariablement grand est la protractée, invariablement petit, la demi-brève, mais un état de plus grand et plus petit doit être considéré par rapport à une brève ou une longue. »

Ce commentaire apporte les précisions qui manquaient : si désigner une figure de note par les termes « grand » ou « petit », alors qu'on a le choix entre quatre valeurs possibles, paraît vague, ordonner strictement les valeurs supprime les ambiguïtés. Ainsi, l'algorithme de construction du *prastāra* prescrit de poser au-dessous d'une figure de note, la figure de note de valeur immédiatement inférieure et, grâce à sa généralité, la règle, telle qu'elle est formulée, peut être utilisée quelles que soient les valeurs rencontrées sur une ligne.

Le sens de lecture n'est pas précisé dès le début, mais il apparaît clairement dans la suite que celle-ci doit avoir lieu de gauche à droite et « la première (plus grande) » est à comprendre comme la première à partir de la gauche. Kallinātha le précise en expliquant ce qu'il faut entendre par *le reste est comme au-dessus* :

[6] *Nyasyālpam ādyān mahato 'dhastāc cheṣaṃ yathopari | prāg ūne vāmasaṃsthāṃs tu saṃbhave mahato likhet | alpān asaṃbhave tālapūrtyai bhūyo 'py ayaṃ vidhiḥ || sarvadrutāvadhiḥ kāryaḥ prastāro 'yaṃ laghau gurau | plute vyaste samaste ca na tu vyaste drute 'sti saḥ* ||

> « Après avoir posé une plus petite au-dessous du premier plus grand composant d'un rythme musical de plusieurs membres, le reste doit être écrit en le plaçant à droite ; on écrira au-dessous, exactement tel que [c'est écrit] sur la ligne du dessus. »

Le fait de placer « à droite » se déduit logiquement de la règle qui indique que s'il faut compléter le rythme — et il le faudra, puisqu'on a diminué une figure de note —, cette complétion s'effectuera sur la gauche : *S'il y a un manque au début, on écrira, si cela est possible, de plus grandes sur la gauche — en cas d'impossibilité des petites — pour compléter le rythme.*

Voici le commentaire de Kallinātha expliquant la manière de compléter :

> « S'il y a au début une partie du mètre musical qui manque, pour compléter le mètre en question, on écrira, quand cela est possible, *des plus grandes*, c'est-à-dire des protractées, des longues ou des brèves ; quand cela est impossible, *des petites*, c'est-à-dire des demi-brèves, placées sur la partie gauche par rapport à ce qui est écrit, à proximité des composants déjà écrits ; telle est la signification. »

Ici encore une précision est apportée pour interpréter les mots « grands » et « petits » : on sait déjà, par l'explication précédente, ce qu'il faut entendre par « grand », mais Kallinātha nous indique cette fois l'ordre dans lequel il faut choisir les figures de notes pour effectuer la complétion : on essaiera d'abord de mettre une protractée, sinon une longue, sinon une brève et si tout cela est impossible une demi-brève. On complète donc le mètre musical en écrivant de droite à gauche les plus grandes figures de notes possible.

Voici sur un exemple le fonctionnement de cet algorithme. Nous écrirons la construction des quatre premières lignes du *prastāra* pour sept *druta*. Les rythmes soulignés sont les rythmes qui apparaîtront dans le *prastāra* ; les nombres entre parenthèses donnent la valeur, en demi-brèves, de la ligne ; un cycle est achevé quand cette valeur est égale à sept.

On commence par mettre une demi-brève et une protractée (1+6) :	**o S̀** (7)
La « première plus grande » est la protractée ; on écrit donc au-dessous, une « plus petite », soit une longue :	**S** (4)
Il n'y a pas de « reste au-dessus », donc rien à écrire sur la droite. Il manque trois demi-brèves « pour compléter le rythme », on écrit donc la valeur de trois demi-brèves, sur la gauche, en commençant par la plus grande valeur possible, soit une brève, et on ajoute, encore sur la gauche, une demi-brève. Fin d'un premier cycle :	**o I S** (7)

On recommence. « La première plus grande » est la brève, on écrit donc une demi-brève au-dessous :	o	(1)
Il y a, cette fois-ci, « un reste au-dessus », la longue, que l'on recopie sur la droite :	o S	(5)
Il manque la valeur de deux demi-brèves, on peut donc écrire une brève sur la gauche pour compléter. Fin d'un deuxième cycle.	I o S	(7)
La « première plus grande » est la brève, on écrit donc une demi-brève au-dessous :	o	(1)
On recopie le « reste du dessus » sur la droite :	o o S	(6)
Il manque une demi-brève que l'on écrit sur la gauche :	o o o S	(7)

Construction du *prastāra*

On poursuit le processus jusqu'à obtenir une ligne qui ne contient que des demi-brèves : *Cette prescription, dont le terme est [un rythme composé] entièrement de demi-brèves, doit être exécutée à plusieurs reprises*. On trouvera page 171 les *prastāra* des mesures de une à sept demi-brèves.

Une première propriété de cet algorithme est de donner exhaustivement la construction de toutes les combinaisons possibles des quatre figures de notes entrant dans la composition d'une mesure musicale de longueur donnée ; en effet, la construction d'une ligne à partir de la précédente se fait en recopiant celle-ci avec un changement a minima d'une seule figure de note : on choisit la valeur immédiatement inférieure. Nous verrons plus loin que l'ordre d'exécution de l'algorithme — on recopie à droite, on complète sur la gauche —, permet d'ordonner les lignes du *prastāra*.

Saṃkhyā

Dénombrement. Śārṅgadeva donne une règle pour compter le nombre des combinaisons obtenues, en fonction du nombre de demi-brèves que comporte le rythme musical pour lequel on a construit le *prastāra*. Voici cette règle :

> *Ayant posé dans l'ordre les nombres un et deux, le dernier est [autant que possible] ajouté aux deuxième, quatrième et sixième nombres qui précèdent — en l'absence du quatrième et du sixième, aux troisième et cinquième nombres — ; on écrira cette somme régulièrement au début. Une suite de nombres placés ensemble vers la droite est ainsi construite et celle-ci est complète avec*

les nombres qui comptent les demi-brèves se trouvant dans un mètre choisi[7].

Le recours au commentaire de Kallinātha permet de mieux comprendre comment établir la suite des nombres qui, comme les *prastāra*, se construit de manière itérative :

> « Après avoir posé le nombre un en premier, on posera, immédiatement après, le nombre deux, à droite en se déplaçant sur une ligne horizontale. On ajoutera, quand cela est possible, le dernier nombre avec le deuxième, le quatrième et le sixième nombre qui se tiennent à gauche par rapport au dernier. Au début, le nombre deux fait office de dernier ; dans ce cas, parce qu'il n'existe que le deuxième parmi les précédents, après avoir fait sa somme avec le nombre un, on écrira le nombre trois sur la droite. Ensuite, après avoir fait la somme de ce [trois] devenu dernier, avec le nombre deux, deuxième parmi les précédents, et — à cause de la phrase : « en l'absence du quatrième et du sixième, ajouté [autant que possible] régulièrement au troisième et cinquième nombres » —, en l'absence de ce quatrième et en substitution de celui-ci, avec le nombre un qui est le troisième ; on écrira le nombre six à la suite du nombre trois. Après avoir ajouté ce [six], devenu le dernier, au nombre trois, qui est le deuxième, et au nombre un qui est le quatrième, on écrira le nombre dix à la suite du nombre six. Puis, après avoir ajouté ce [dix], dernier, avec le nombre six et le nombre deux, qui sont le deuxième et le quatrième parmi les précédents, et aussi avec le nombre un qui est le cinquième, en substitution du sixième qui manque, on écrira le nombre dix-neuf à la suite du nombre dix. Ensuite, après avoir ajouté ce [dix-neuf], dernier, aux nombres dix, trois et un qui sont les deuxième, quatrième et sixième précédents, on écrira le nombre trente-trois. De cette manière, une suite de nombres, placés ensemble sur la droite est construite. »

Nous avons là, la construction d'une suite récurrente : on calcule les termes, les uns après les autres, en combinant — dans ce cas en additionnant — des nombres précédemment calculés. Pour obtenir un nouvel élément (le rang n) de cette suite, il faut additionner le dernier nombre obtenu (rang $n-1$) avec l'avant-dernier (rang $n-2$), le quatrième à partir du dernier (rang $n-4$) et le sixième, toujours à partir du dernier (rang $n-6$). Il faut donc disposer de six nombres avant de pouvoir commencer à construire cette suite de manière itérative, c'est le sens des exceptions formulées dans la règle : si le quatrième n'existe pas encore, on lui substitue le troisième et si le sixième n'existe pas, on le remplace par le cinquième. Comme les deux premiers termes de la suite, 1 et 2, nous sont donnés, ces exceptions ne concernent

[7] *Ekadvyaṅkau kramān nyasya yuñjītāntyaṃ purātanaiḥ | dvitīyaturyaṣaṣṭhāṅkair abhave turyaṣaṣṭhayoḥ ‖ tṛtīyapañcamāṅkābhyāṃ kramāt taṃ yogam agrataḥ | likhed dakṣiṇasaṃsthaivam aṅkaśreṇī vidhīyate ‖ sā cāṅkair iṣṭatālasthadrutasaṃkhyaiḥ samāpyate ‖*

que les rangs trois à six ; au delà, la règle générale s'applique, que l'on note aujourd'hui de la manière suivante :

$$u_n = u_{n-1} + u_{n-2} + u_{n-4} + u_{n-6}$$

où u_n est le n-ième terme de la suite.

Les nombres de cette suite comptent le nombre de combinaisons possibles des quatre figures de notes dans un rythme musical de longueur fixe ; chaque nombre correspond à un rythme et son rang compte le nombre de demi-brèves mesurant la longueur du rythme, ainsi que nous le dit Śārṅgadeva :

> *Les variétés de rythmes sont comptées par les nombres placés ici, augmentés régulièrement à chaque demi-brève, c'est-à-dire : une demi-brève, une brève, une more et demie, une longue, deux mores et demie, une protractée, trois mores et demie*[8] .

La construction de cette suite est illustrée dans le tableau suivant. Les termes de la suite sont sur la diagonale du tableau ; on reconnaît la suite de nombres citée au début de cet article.

Les colonnes sont numérotées par le rang des termes de la suite, donc par le nombre de demi-brèves mesurant la longueur d'un rythme musical : il y a dix variations pour un rythme d'une longueur de cinq demi-brèves. Les lignes montrent le calcul de chaque terme de la suite ; par exemple, le terme de rang six, 19, est obtenu en faisant la somme des termes de rang un, deux, quatre et cinq.

Rang :	1		2		3		4		5		6		7		8
	1														
	1		2												
	1	+	2	=	3										
	1	+	2	+	3	=	6								
	1			+	3	+	6	=	10						
	1	+	2			+	6	+	10	=	19				
	1			+	3			+	10	+	19	=	33		
			2			+	6			+	19	+	33	=	60

Le poème mnémotechnique trouve maintenant son explication, à la lumière de la règle donnée par Śārṅgadeva : « Les combinaisons jusqu'à six

[8] *Druto laghuḥ sārdhamātro guruḥ sārdhadvimātrikaḥ* | *plutaḥ sārdhatrimātraś cety ekaikadrutavardhitaiḥ* | *tālabhedāḥ kramād aṅkaiḥ saṃkhyāyante sthitair iha* ||

demi-brèves se mémorisent par : un, deux, trois, six, dix, dix-neuf » ; les six premiers termes de la suite nous sont donnés. « Les deux derniers additionnés et aussi les deux [séparés] par un intervalle, on aura aussi le dénombrement avec la somme des uns et des autres ». Cette partie nous permet de construire les termes suivants en additionnant le dernier terme calculé avec l'avant-dernier et deux autres termes précédents, mais en sautant un terme intermédiaire à chaque fois, c'est-à-dire les quatrième et sixième termes à partir du dernier.

Explications

La construction de la suite

Le problème posé est un intéressant, et difficile, problème de combinatoire et la solution donnée à ce problème par Śārṅgadeva est très séduisante par sa simplicité : tout repose sur la construction du *prastāra*.

Le problème, vu sous l'angle musical, consiste à trouver toutes les variations rythmiques possibles dans un temps fixé, en utilisant quatre figures de notes, ayant entre elles les rapports de durée : un, deux, quatre et six.

On peut donner une formulation mathématique de ce problème : trouver toutes les manières possibles d'écrire un nombre entier comme la somme des nombres un, deux, quatre et six. Par exemple : $7 = 1 + 6 = 2 + 2 + 2 + 1 = 4 + 2 + 1$, etc. C'est un problème complexe de théorie des nombres qui a été développé et étudié par Srinivas Ramanujan et G. H. Hardy au début du XXe siècle ; on l'appelle la partition des entiers. C'est aussi, de nos jours, un problème de programmation informatique qui se pose pour les automates qui rendent la monnaie : comment rendre au mieux une certaine somme d'argent en fonction des pièces disponibles dans la machine.

Le point de vue musical ajoute une contrainte supplémentaire : l'ordre est signifiant. En effet, pour un mathématicien, $7 = 4 + 2 + 1 = 1 + 4 + 2 = 2 + 4 + 1 = \cdots$ indifféremment. Pour un musicien, les rythmes : **SIo** (𝅗𝅥 ♩ ♪) ou **oSI** (♪ 𝅗𝅥 ♩) ou **ISo** (♩ 𝅗𝅥 ♪) sont différents bien que de même durée. Cette contrainte n'apparaîtra pas dans le calcul car, du fait de la méthode itérative utilisée, une fois correctement établis les six premiers termes de la suite, le principe de récursivité va répercuter les résultats de cette contrainte à tous les termes de la suite.

Pour comprendre la construction de la suite, il faut se reporter au tableau page 171, donnant les *prastāra* construits pour des mesures de une à sept demi-brèves. En lisant chaque tableau de bas en haut et de droite à gauche, on constate que la dernière colonne comporte, à partir du bas, des lignes finissant par une demi-brève, valeur 1, puis viennent des lignes finissant par une brève, valeur 2, puis des lignes finissant par une longue, valeur 4, enfin des lignes finissant par une protractée, valeur 6. L'algorithme de construction des *prastāra* donne, non seulement une construction exhaustive, mais ordonne aussi les lignes de cette construction en les regroupant par finale identique. Nous verrons plus loin une description de cet ordre.

Si l'on adopte le point de vue mathématique pour lire les *prastāra*, on pourrait remplacer chaque figure de note par sa valeur ; on écrirait alors : 2, 1, 4, à la place de **I o S**, et on lirait : « la ligne 2, 1, 4 est une partition de l'entier 7 », c'est-à-dire $7 = 2 + 1 + 4$. Si on décrit, par exemple, le *prastāra* pour sept demi-brèves, en adoptant ce point de vue on dira alors : de bas en haut, viennent d'abord les partitions de l'entier 7 dont la somme se termine par 1 et il y en a autant que de partitions de l'entier $7 - 1 = 6$. En effet, une partition quelconque de 7 se terminant par 1, s'écrit : $7 = x + \cdots + 1$, pour les avoir toutes, il suffit de connaître toutes les partitions de $7 - 1 = x + \cdots$

Puis, viennent les partitions de l'entier 7 se terminant par 2 et, suivant le même raisonnement que pour celles se terminant par 1, il y en a autant que les partitions de l'entier $7 - 2 = 5$; puis les partitions de 7 finissant par 4, qui sont aussi nombreuses que celles de l'entier $7 - 4 = 3$; enfin, les partitions se terminant par 6, qui sont autant que celles de $7 - 6 = 1$. En fin de compte, le nombre de partitions de l'entier 7 est égal à la somme des nombres de partitions des entiers 6, 5, 3 et 1, soit : $u_7 = u_6 + u_5 + u_3 + u_1$. Ou, généralement, le nombre u_n de partitions de l'entier n est égal à :

$$u_n = u_{n-1} + u_{n-2} + u_{n-4} + u_{n-6}$$

C'est-à-dire la formule de récurrence donnée par Śārṅgadeva.

Le tableau page 172 est une illustration de la construction de la suite, vue sous l'angle mathématique : pour le *prastāra* de sept demi-brèves, on a noté, de bas en haut, à l'aide d'une double flèche verticale l'ensemble des lignes se terminant par une demi-brève, valeur 1, puis celles se terminant par une brève, valeur 2, etc. et on a mis en valeur, par couleurs identiques, les *prastāra* précédents qui entrent dans sa composition, quand on « oublie » la dernière colonne, soit les *prastāra* de six, cinq, trois et une demi-brèves.

La propriété des *prastāra* que nous avons utilisée pour montrer comment dénombrer les combinaisons de figures de notes n'a pas échappé à Śārṅgadeva, qui écrit en conclusion de cette section :

> *À partir [des termes] de la somme des nombres dont on obtient le dernier nombre, on compte avec ceux-ci, successivement à partir de la fin, les variétés qui finissent par une demi-brève, une brève, une longue et une protractée ; tels sont les dénombrements enseignés par le musicien*[9].

Ordre sur les *prastāra*

Nous avons remarqué, pour expliquer la construction de la suite, que la dernière colonne des *prastāra* était rangée, de bas en haut, par ordre croissant des valeurs des figures de notes. En fait, tout *prastāra* est ordonné, par sa construction même, de bas en haut et de droite à gauche, selon l'ordre lexicographique.

C'est l'ordre de classement des mots dans un dictionnaire, il permet de comparer les mots lettre par lettre en utilisant un ordre « naturel » défini sur l'ensemble des lettres : l'ordre alphabétique. On peut le définir de la manière suivante pour l'alphabet latin : un mot $\mathcal{M}_1$ — une séquence de lettres — est placé après un mot $\mathcal{M}_2$ si :

- il y a une lettre dans $\mathcal{M}_2$ postérieure à celle qui occupe la même position dans $\mathcal{M}_1$
- les lettres avant cette position sont identiques dans $\mathcal{M}_1$ et $\mathcal{M}_2$
- les lettres qui suivent cette position n'interviennent pas dans le classement

On a, par exemple, le classement suivant : alpha, alphabet, alphabétique, alphabétiquement, alphabétisation, alphabétisme...

Dans le cas présent, l'« alphabet » a quatre lettres, classées dans l'ordre : **o**, **I**, **S**, **S̀** et on peut constater, en lisant le *prastāra* pour sept demi-brèves (page 171) de droite à gauche et de bas en haut, que les lignes se classent bien selon l'ordre lexicographique :

$$\mathbf{ooooooo} \prec \mathbf{Iooooo} \prec \cdots \prec \mathbf{oooooI} \prec \cdots \prec \mathbf{oooS} \prec \cdots \prec \mathbf{o\grave{S}}$$

C'est là une propriété qui résulte de l'ingénieux algorithme de construction du *prastāra* et qui a son application par la suite dans la procédure appelée « *naṣṭa* » (disparu). Cette opération consiste à reconstituer une ligne effacée dans le *prastāra*, à l'aide de la connaissance de son seul rang parmi les lignes. L'ordre lexicographique joue alors un rôle essentiel, car c'est un

[9] *Yadaṅkayogād antyo 'ṅko labdhas tair antataḥ kramāt* | *bhedā drutāntalaghvantagurvantāś ca plutāntakāḥ* | *saṃkhyāyanta iti proktāḥ saṃkhyā niḥśaṅkasūriṇā* ||

ordre « total » : quels que soient les mots, on peut toujours les classer sans ambiguïté. Nous ne développerons pas plus avant cette procédure complexe dans cet article.

Génie mathématique

Cet algorithme est une bonne illustration des méthodes employées par les mathématiciens indiens pour résoudre des problèmes, souvent complexes, dont la solution est donnée sous forme itérative. Ce traité de musique, le *Saṃgītaratnākara*, en présente plusieurs exemples dont l'ingéniosité est réjouissante pour qui est sensible à une certaine esthétique mathématique. Les techniques sous-jacentes sont parfois sophistiquées comme l'usage d'une double suite récurrente permettant de compter le nombre de figures de notes — par exemple le nombre de demi-brèves — utilisées dans un *prastāra*. Nous donnerons cette construction comme un exemple supplémentaire de l'esprit d'analyse et d'abstraction dont font preuve les *paṇḍits* indiens.

Si l'on observe le *prastāra* pour sept demi-brèves (page 172), nous avons vu qu'il était construit comme l'« empilement » des *prastāra* pour six, cinq, trois et une demi-brèves, bordé sur la droite par une dernière colonne, ordonnée de bas en haut par ordre croissant des valeurs des figures de notes qu'elle contient.

De plus, pour chaque ensemble de valeurs de cette dernière colonne, les figures de notes sont en nombre égal au nombre de lignes du *prastāra* en regard duquel elles se trouvent : il y a, dans la dernière colonne, autant de demi-brèves que de lignes dans le *prastāra* pour six demi-brèves, autant de brèves que de lignes dans le *prastāra* pour cinq demi-brèves, etc. Si donc, on veut connaître le nombre de demi-brèves utilisées dans le septième *prastāra*, il suffit d'additionner le nombre de demi-brèves utilisées dans les *prastāra* pour six, cinq, trois et une demi-brèves et d'ajouter le nombre de demi-brèves contenues dans la dernière colonne, c'est-à-dire le nombre de lignes du *prastāra* pour six demi-brèves.

Si l'on note v_7 ce nombre de demi-brèves, il est donc égal à :

$$v_7 = v_1 + v_3 + v_5 + v_6 + u_6$$

où le dernier terme, u_6, est le sixième terme de la suite « *saṃkhyā* » que nous avons étudiée plus haut. Śārṅgadeva nous donne une règle, toujours de manière itérative, permettant de construire une nouvelle suite, parallèlement à la suite *saṃkhyā*, qui comptera, d'une manière générale, le nombre de demi-brèves d'un *prastāra* arbitraire[10]. Il fait également remarquer que cette même suite permet de compter le nombre de brèves, longues et protractées en « décalant » les rangs : si le dernier nombre calculé compte les demi-brèves, l'avant-dernier compte les brèves, le quatrième précédent, les longues et le sixième précédent, les protractées.

Le *Saṃgītarātnakara* nous montre bien d'autres constructions faisant appel à un bagage mathématique important, non seulement dans le chapitre consacré au rythme, mais aussi, par exemple, dans le chapitre sur la gamme où apparaissent des tableaux *(khaṇḍameru)* calculant les combinaisons des modulations dans les systèmes de tonalité. Cela laisse entrevoir un usage des mathématiques qui s'éloigne de sa perception habituelle comme discipline au service des simples calculs de transactions, ou comme outil subordonné à l'astronomie, pour en faire un moyen de modélisation et d'exploration théorique pour d'autres disciplines — dans notre cas, la musique — sans doute bien avant le XIVe siècle.

[10]Transposée dans nos notations mathématiques contemporaines, la construction de Śārṅgadeva est la suivante : on a deux suites récurrentes parallèles : u_n est la suite « *saṃkhyā* » que nous avons étudiée, et v_n compte le nombre de demi-brèves entrant dans la composition du n-ième *prastāra* : $\begin{cases} u_n & = & u_{n-1} + u_{n-2} + u_{n-4} + u_{n-6} \\ v_n & = & u_{n-1} + v_{n-1} + v_{n-2} + v_{n-4} + v_{n-6} \end{cases}$

						o S̀
						o I S
						I o S
						o o o S
						o S I
						o I I I
						I o I I
						o o o I I
						S o I
						I I o I
						o o I o I
						o I o o I
						I o o o I
						o o o o o I
					S̀	S̀ o
					I S	I S o
					o o S	o o S o
					S I	S I o
					I I I	I I I o
					o o I I	o o I I o
					o I o I	o I o I o
					I o o I	I o o I o
					o o o o I	o o o o I o
				o S	o S o	o S o o
				o I I	o I I o	o I I o o
				I o I	I o I o	I o I o o
				o o o I	o o o I o	o o o I o o
			S	S o	S o o	S o o o
			I I	I I o	I I o o	I I o o o
			o o I	o o I o	o o I o o	o o I o o o
		o I	o I o	o I o o	o I o o o	o I o o o o
	I	I o	I o o	I o o o	I o o o o	I o o o o o
o	o o	o o o	o o o o	o o o o o	o o o o o o	o o o o o o o
1	2	3	4	5	6	7

TAB. 1 – Les *prastāra* pour une à sept *druta*

$$u_7 = u_6 + u_5 + u_3 + u_1$$

$$u_n = u_{n-1} + u_{n-2} + u_{n-4} + u_{n-6}$$

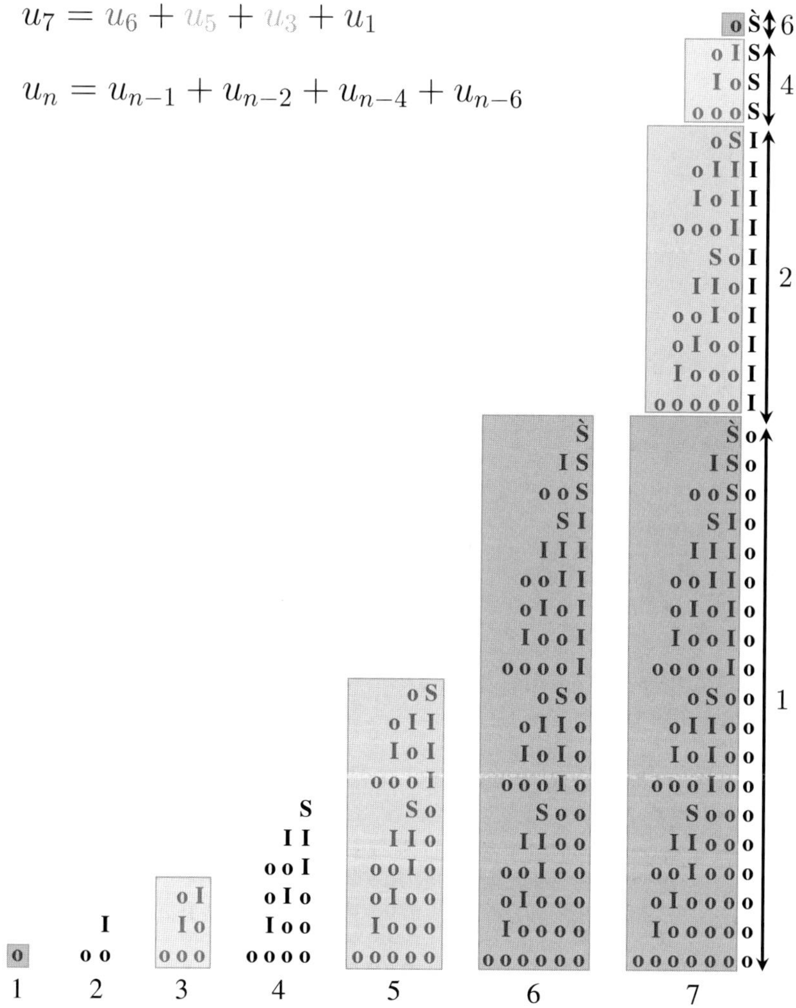

TAB. 2 – Construction de la suite *saṃkhyā*

Références

Textes sanskrits

KEDĀRABHAṬṬA
1890 *Vṛttaratnākara*. Bombay : The Nirṇaya-Sâgara Press. Avec le commentaire de Nârayâṇabhatta.

NIJENHUIS, Emmie Te, éditeur
1992 *Saṃgītaśiromaṇi : a medieval handbook of Indian music*. Leiden : E.J. Brill.

PIṄGALA
1908 *The Chhanda Shâstra*. Bombay : The Nirnaya-Sāgara Press. Avec le commentaire *Mṛtasanjivanī* de Halāyudha Bhaṭṭa. Édité par Paṇdita Kedāranātha and Wāsudava Laxmaṇ Shāstrī Paṇashīkar.

ŚĀRṄGADEVA
1896 *Saṃgītaratnākara*. Punyākhyapattana : Ānandāśramamudraṇālaya (Ānandāśramaṃskṛtagranthāvaliḥ). Avec le commentaire *Kalānidhi* de Kallinātha.
1951 *Saṅgītaratnākara*. (Édité par SASTRI, Pandit S. Subrahmanya). Adyar : The Adyar Library (The Adyar Library Series — N° 78). Avec les commentaires *Kalānidhi* de Kallinātha et *Sudhākara* de Siṁhabhūpāla.

Textes mathématiques

BÓNA, Miklós
2002 *A Walk Through Combinatorics*. Singapore : World Scientific Publisher Co. Pte. Ltd.

HARDY, Godfrey Harold
1999 *Ramanujan : twelve lectures on subjects suggested by his life and work*. USA : American Mathematical Society. Première édition, Cambridge, 1940.

KULKARNI, Amba
2008 « Recursion and Combinatorial Mathematics in Chandashāstra », *http ://arxiv.org/abs/math/0703658*.

LEHMER, D. H.
1970 « Permutations with strongly restricted displacements », dans *Combinatorial Theory and its Applications*. volume II, Proc. Colloq., Balatonfured, 1969. Amsterdam : North Holland, p. 755–770.

WILF, Herbert S.
2006 *generatingfunctionology*. Wellesley : A.K. Peters, Ltd.

ZOGHBIU, Antoine et STOJMENOVIĆ, Ivan
1998 « Fast Algorithms For Generating Integer Partitions », *International Journal of Computer Mathematics*, 70, p. 319–332.

Mādhava series for π and its fast convergent approximations

K. Mahesh, Venketeswara Pai R. and K. Ramasubramanian
Cell for Indian Science and Technology in Sanskrit
Department of Humanities and Social Sciences, IIT Bombay
Powai, Mumbai, India

Abstract

Way back in the 14th century, the mathematicians and astronomers residing on the banks of the river Nila in the south Malabar region of Kerala—stumbling upon the problem of finding the exact relationship between the arc and the corresponding chord of a circle, and certain other problems associated with that—came very close to inventing what goes by the name of infinitesimal calculus today. They had advanced to the point of discovering the series expansions for π as well as the sine and the cosine functions. In fact, it is for the first time that we come across a mathematician, by name Mādhava (c. 1340–1420 CE), departing from the beaten track of using the standard polygon-doubling method for evaluating π, and presenting an infinite series (referred to as Mādhava series in the title of the paper), that is popularly known as Gregory-Leibniz series. Since this is an excrutiatingly slowly converging serie, it is of hardly any value as a numerial approximator to π. To circumvent this problem, Mādhava invented brilliant algorithms, that are not only computationally far more efficient, but also involve sophisticated mathematical analysis.

1 Introduction

The relation between the circumference of a circle and its diameter—that goes by the name *paridhi-vyāsa-sambandha* in Sanskrit—has been expressed in various forms by several brilliant minds from the period of *Śulva-sūtras* (before 8th century BCE). Of the different approaches taken by the ancients to obtain this relation, the geometric approach which involves inscribing/circumscribing a polygon inside/outside a circle and then estimating the circumference of the circle by calculating the polygon's perimeter, seems to be the oldest one. A historical survey of the contributions made by the three major traditions—Hellenistic, Chinese and Indian—indicates that perhaps this was the only approach known to mathematicians of the antiquity.

It is quite interesting to note that the first departure from the beaten track of using the standard polygon-doubling method for evaluating π seems to have been done by Mādhava of Saṅgamagrāma. Besides making this revolutionary departure by way of presenting an expression for the exact value of π in the form of an infinite series (the so called Gregory-Leibniz series for $\frac{\pi}{4}$), Mādhava also seems to have systematically developed algorithms for obtaining several fast convergent approximations to π. As will be seen below, these algorithms are not only computationally far more efficient, but also involve sophisticated mathematical analysis.

One of the algorithms that goes by the name *antya-saṃskāra*, transforms a slowly converging series into fast convergent ones [11], by applying certain techniques that are described as error-minimization techniques in modern parlance. In the present paper, we intend to present the results of our study (theoretical as well as computational) of this *antya-saṃskāra* algorithm that has been described by Śaṅkara Vāriyar (c. 1500–1560) in his commentary *Kriyākramakarī* [8, pp.386–390]—a detailed exposition on the *Līlāvati* of Bhāskarācārya. It may be mentioned that these algorithms have also been explained by Jyeṣṭhadeva in his *Yuktibhāṣā* [6].

2 Approximations to π: A brief historical survey

Archimedes of Syracuse (3rd century BCE) in his treatise on the *Measurement of a Circle* gives an estimate of the ratio of the circumference to the diameter of a circle (the value of π) by specifying a lower and an upper bound. According to the third proposition of this treatise [4]:

$$3\frac{10}{71} < \pi < 3\frac{1}{7}. \tag{1}$$

It has been shown by Edwards [5] that Archimedes must have arrived at these bounds by considering the inscribed and the circumscribed polygons of sides 48 and 96 respectively. The average value of these bounds turns out to be 3.141836, which is correct to two decimal places. About 150 CE, Ptolemy of Alexandria in his *Almagest* uses a value to π which in sexagesimal units is represented by 3;8,30, whose rational approximation and decimal equivalents (to four digits) are $\frac{377}{120}$ and 3.1417 respectively. Then commences the slumber period in the west which lasts for almost 13 centuries till around the end of 15th century—when Francois Viete, Galileo Galilei, Bonaventura Cavalieri, etc., appear—which marks the beginning of Renaissance period.

One of the mathematicians Viete, belonging this period, in his *Canon mathematicus*—after pleading for the use of decimal system of representation rather than sexagesimal—specifies the ratio of the circumference to the diameter of a circle. According to Viete, the semi-circumference of a circle whose diameter is $200{,}000,\frac{000.00}{}$, is $314{,}159\frac{265.35}{1{,}000.00}$ [2]. This amounts to specifying the value of π correct to almost 10 decimal places.[1] We may conclude our discussion on the contribution of the West at this stage, by simply mentioning the fact that Viete too obtained this value by polygon-doubling method as Archimedes.

Coming to the Chinese tradition, according to Needham [7], Liu Hui (4th century CE) starting with a polygon of 192 sides and then moving on to the one with 3,072 sides seems to have obtained π value correct to 5 decimal places. Zu Chongzhi an astronomer, mathematician and engineer of the later half of the 5th century, based on the earlier writings as well as his on computations, seems to have recommended the use of $\frac{355}{113}$ and $\frac{22}{7}$ as close and rough approximations for π. Needham also notes that Chao Yu-Chhin of 14th century, had obtained π value correct to 6 decimal places essentially following the same technique of polygon-doubling, but choosing a polygon of 16,384 sides.

As far as the Indian tradition goes, the earliest reference to π occurs in a class of texts known as *Śulva-sūtras*. These texts do not present the value of π explicitly. However, in the context of discussing methods for constructing altars of different shapes with the constraint that they should have same

[1] Editor's note: 3.1415926535 is actually correct to 10 decimal places, but the following decimal being 9, 3.1415926536 is nearer to the exact value of π.

area—a problem which goes by the name of measure preserving transformations today—one finds the value of π mentioned implicitly. From the formula for the radius of the circle given in the *Śulva-sūtra* text, in terms of the length of the side of the square in order to make this transformation, it is straightforward to deduce the value of π. If the side of the square that needs to be transformed into a circle is $2a$, then according to the text the radius of the circle r is [13]:

$$r = a + \frac{1}{3}(a\sqrt{2} - a) = \frac{a}{3}(2 + \sqrt{2}). \qquad (2)$$

The constraint that the transformation has to be area preserving translates to the condition $\pi r^2 = 4a^2$. Hence we have the relation

$$\pi \left[\frac{1}{3}(2 + \sqrt{2})\right]^2 = 4 \qquad \text{or} \qquad \pi \approx 3.0883, \qquad (3)$$

which is correct only to one decimal place. Āryabhaṭa in his *Āryabhaṭīya* (c. 499 CE) having specified the ratio of the circumference to the diameter of a circle as [1]

$$\frac{\textit{pariṇāha}\text{ (circum.)}}{\textit{viṣkambha}\text{ (dia.)}} = \pi = \frac{62832}{20000} = 3.1416, \qquad (4)$$

explicity refers to the value as 'approximate' (*āsanna*). The above value is correct to 4 decimal places. Among several other interesting approximations to π given by the Indian astronomers of the later period, the ones which are of particular interest to us, as far as this paper is concerned, are those based upon the algorithms presented by Kerala mathematicians.

A verse attributed to Mādhava (c. 1340–1420)—the pioneer of the Kerala School of mathematicians—by Śaṅkara Vāriyar in his *Kriyākramakarī* gives a rational approximation to π that is correct to 11 decimal places [8, p. 377].

> *vibudhanetragajāhihutāśanatriguṇavedabhavāraṇabāhavaḥ* |
> *navanikharvamite vṛtivistare paridhimānamidaṃ jagadurbudhāḥ* ||

The first half of the above verse presents a 13 digit number in what is known as *Bhūtasaṅkhyā* system,[2] and this forms the numerator of the rational approximation to π. The word *nikharva* appearing in the second line represents

[2]According to this system of representing numbers, *Vibudha*=33, *Netra*=2, *Gaja*=8, *Ahi*=8, *Hutāśana*=3, *Tri*=3, *Guṇa*=3, *Veda*=4, *Bha*=27, *Vāraṇa*=8, *Bāhu*=2, *Navanikharva*=9×10^{11}.

the number 10^{11}. Thus according to the above verse

$$\pi = \frac{2827433388233}{9 \times 10^{11}} = 3.141592653592. \qquad (5)$$

Since obtaining π values to such high degree of precision cannot simply happen by chance, it is imperative upon us to conclude that Mādhava must have had some kind of algorithm to arrive at such precise values. A survey of the works of the Kerala mathematicians reveals that Mādhava not only had one, but many algorithms. However, none of the works of Mādhava on mathematics—if any were composed by him—are extant today. We come to know of the brilliant workings of Mādhava's mind only through the citations of his verses by later astronomers.

3 Mādhava series for π

The infinite series for $\frac{\pi}{4}$ is enunciated by Mādhava in the form of a short verse composed in *āryā* metre. This verse is cited by Śaṅkara Vāriyar in his commentary *Kriyākramakarī* on *Līlāvatī* [8, p. 379]:

> *vyāse vāridhinihate rūpahṛte vyāsasāgarābhihate* |
> *triśarādiviṣamasaṅkhyābhaktamṛṇaṃ svaṃ pṛthak kramāt kuryāt* ||
>
> The diameter multiplied by four and divided by unity [is found and saved]. Again the products of the diameter and four are divided by the odd numbers like three, five, etc., and the results are subtracted and added in order.

The series encoded in the above verse is the well known series

$$\frac{\pi}{4} = 1 - \frac{1}{3} + \frac{1}{5} - \frac{1}{7} \ldots, \qquad (6a)$$

which may also be written as

$$C = 4D\left(1 - \frac{1}{3} + \frac{1}{5} - \ldots\right), \qquad (6b)$$

where C is the circumference of the circle whose diameter is D. We shall now present the derivation of the above series as outlined in *Yuktibhāṣā* [6, pp. 183–98]. For this, let us consider the quadrant OP_0P_nS of the square circumscribing the given circle (see Figure 1) of radius r. Divide the side P_0P_n into n equal parts (n very large). The resulting segments P_0P_i's ($i = 1, 2, \ldots, n$) are known as the *bhujās* and the line joining their extremities to the centre, or OP_i's, are known as *karṇas*. The points of intersection of these *karṇas* and the circle are denoted by A_i. The *bhujās* P_0P_i, the *karṇas* $k_i = OP_i$ and the east-west line OP_0 form right-angled triangles whose hypotenuses are given by

$$k_i^2 = r^2 + \left(\frac{ir}{n}\right)^2. \tag{7}$$

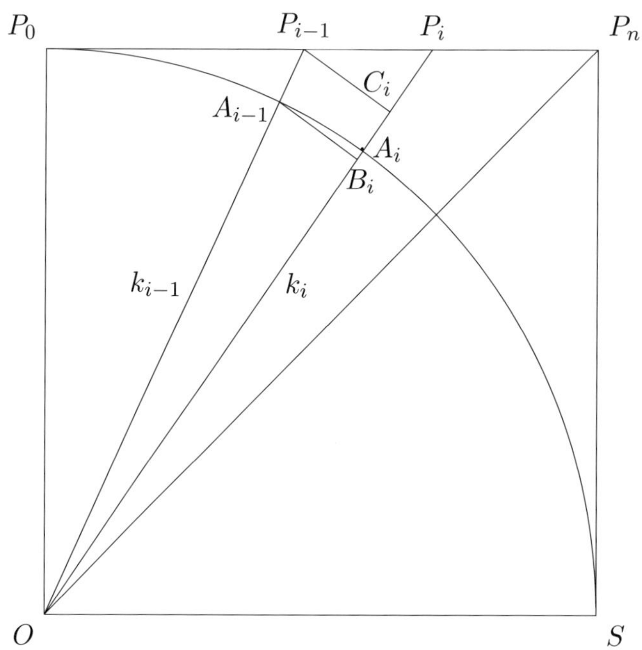

Figure 1: Geometrical construction used in the proof of the infinite series for π.

Considering two successive *karṇas*—ith and the previous one as shown in the figure—there are two pairs of similar triangles, $OP_{i-1}C_i$; $OA_{i-1}B_i$ on the one hand, and $P_{i-1}C_iP_i$; P_0OP_i on the other hand. It can be shown that

$$A_{i-1}B_i = \left(\frac{r}{n}\right)\left(\frac{r^2}{k_{i-1}k_i}\right). \tag{8}$$

Now the text presents the crucial argument: when n is large, the Rsines $A_{i-1}B_i$ corresponding to different arc-bits $A_{i-1}A_i$ can be taken as the arc-bits themselves. Thus, $\frac{1}{8}$th of the circumference of the circle can be written as the sum of the contributions given by (8).

$$\frac{C}{8} \approx \left(\frac{r}{n}\right)\left[\left(\frac{r^2}{k_0 k_1}\right) + \left(\frac{r^2}{k_1 k_2}\right) + \cdots + \left(\frac{r^2}{k_{n-1} k_n}\right)\right]. \tag{9}$$

It is further argued in the text that the denominators $k_{i-1}k_i$ may be replaced by the square of either of the *karṇas* i.e., by k_{i-1}^2 or k_i^2 since the difference is negligible. Thus (9) may be re-written in the form

$$\begin{aligned} \frac{C}{8} &= \sum_{i=1}^{n} \frac{r}{n}\left(\frac{r^2}{k_i^2}\right) \\ &= \sum_{i=1}^{n} \left(\frac{r}{n}\right)\left(\frac{r^2}{r^2 + \left(\frac{ir}{n}\right)^2}\right) \\ &= \sum_{i=1}^{n} \left[\frac{r}{n} - \frac{r}{n}\left(\frac{\left(\frac{ir}{n}\right)^2}{r^2}\right) + \frac{r}{n}\left(\frac{\left(\frac{ir}{n}\right)^2}{r^2}\right)^2 - \ldots\right] \end{aligned} \tag{10}$$

In the series expression for the circumference given above, factoring out the powers of $\frac{r}{n}$, the summations involved are that of even powers of the natural numbers. It was known to the Kerala mathematicians [6, p. 196] that for large n

$$\sum_{i=1}^{n} i^k \approx \frac{n^{k+1}}{k+1}. \tag{11}$$

Using the above relation in (10), we arrive at the result[3]

$$\frac{C}{8} = r\left(1 - \frac{1}{3} + \frac{1}{5} - \frac{1}{7} + \cdots\right), \tag{12}$$

which is same as (6*b*), the well known series for $\frac{\pi}{4}$.

[3] In modern terminology, the above derivation amounts to the evaluation of the following integral

$$\frac{C}{8} = \lim_{n\to\infty} \sum_{i=1}^{n} \left(\frac{r}{n}\right)\left(\frac{r^2}{r^2 + \left(\frac{ir}{n}\right)^2}\right) = r\int_0^1 \frac{dx}{1+x^2}.$$

4 *Antyasaṃskāra*: Estimating Mādhava series by applying end-correction

It is well known that the Mādhava series given above (6*b*) is an extremely slowly converging series—so much so that it requires millions of terms to be considered even to get an accuracy upto four decimal places. This problem has been brilliantly solved by Mādhava by applying a certain technique, which may be called *antyasaṃskāra*, in which the series is rewritten in the following form:

$$C \approx 4D\left(S_p + \frac{1}{a_p}\right). \tag{13}$$

Here S_p represents the sum of the terms in the series that is terminated at $\frac{1}{p}$ (p being an odd number of our choice) and the term $\frac{1}{a_p}$ is a rational approximation to the remaining terms in the series. The application of this term is known as *antyasaṃskāra*. In the discussion that follows this term will simply be referred to as end-correction. The nomenclature stems from the fact that a correction (*saṃskāra*) is applied towards the end (*anta*) of the series, when it is terminated after a certain number of terms.

Three successive approximations to the end-correction ($\frac{1}{a_p}$) have been given by Mādhava, and they may be expressed as:

$$\frac{1}{a_p(1)} = \frac{1}{2(p+1)} \tag{14}$$

$$\frac{1}{a_p(2)} = \frac{\left(\frac{p+1}{2}\right)}{(p+1)^2+1} \tag{15}$$

$$\frac{1}{a_p(3)} = \frac{\left(\frac{p+1}{2}\right)^2+1}{\left[\left(\left(\frac{p+1}{2}\right)^2+1\right)4+1\right]\left(\frac{p+1}{2}\right)}, \tag{16}$$

where as mentioned earlier, p represents the last odd number used as the divisor, at which the series was terminated. The end-correction represented by (15), is given in the following verses [8, p. 379]:

> *yatsaṅkhyayā'tra haraṇe kṛte nivṛttā hṛtistu jāmitayā* |
> *tasyā ūrdhvagatā yā samasaṅkhyā taddalaṃ guṇo'nte syāt* ||
>
> *tadvargo rūpayuto hāro vyāsābdhighātataḥ prāgvat* |
> *tābhyāmāptaṃ svamṛṇe kṛte dhane śodhanañca karaṇīyam* ||

> Half of the succeeding even number, at whichever [odd] number the process of division is terminated, because of boredom [caused by the slow convergence] is the multiplier. The square of that [even number] added by unity is the divisor. The result obtained from these two multiplied by the product of the diameter and four as earlier [and] has to be added if the earlier term [in the series] has been subtracted; subtracted if the earlier term has been added.

The series obtained by applying the end-correction given in the above verses may be written as

$$C \approx 4D \left(1 - \frac{1}{3} + \frac{1}{5} - \ldots + (-1)^{\frac{p-1}{2}} \frac{1}{p} + (-1)^{\frac{p+1}{2}} \frac{\left(\frac{p+1}{2}\right)}{(p+1)^2+1} \right) \tag{17}$$

We now proceed to explain the rationale behind the end-correction given in (17). Suppose we terminate the series given by (6*b*) after the term $\frac{1}{p}$, where p is an odd number of our choice, and consider applying the end-correction (*antyasaṃskāra*) $\frac{1}{a_p}$, as a substitute of the remaining terms in the series, then the series becomes

$$C \approx 4D \left(1 - \frac{1}{3} + \frac{1}{5} \ldots + (-1)^{\frac{p-1}{2}} \frac{1}{p} + (-1)^{\frac{p+1}{2}} \frac{1}{a_p} \right) \tag{18}$$

Now the question arises as to what gives the necessary licence to terminate the series at $\frac{1}{p}$ and represent the rest of the infinite term by a single term (*antyasaṃskāra*)? And more importantly, what is the guarantee that the end-correction gives the right estimate of the remainder terms in the series?

The argument adduced in both *Yuktibhāṣā* and *Kriyākramakarī* in favor of terminating the series at any desired term, still ensuring the accuracy, may be summarized as follows. Let the series for C be written as

$$C = 4D \left(1 - \frac{1}{3} + \frac{1}{5} \ldots + (-1)^{\frac{p-3}{2}} \frac{1}{p-2} + (-1)^{\frac{p-1}{2}} \frac{1}{a_{p-2}} \right), \tag{19}$$

where $\frac{1}{a_{p-2}}$ is the end-correction applied after odd denominator $p-2$. On the other hand, if the end-correction $\frac{1}{a_p}$, is applied after the odd denominator p, then

$$C = 4D \left(1 - \frac{1}{3} + \frac{1}{5} \ldots + (-1)^{\frac{p-1}{2}} \frac{1}{p} + (-1)^{\frac{p+1}{2}} \frac{1}{a_p} \right). \tag{20}$$

If the end-corrections chosen were exact, viz., they happen to be the right estimate of the remainder terms in the series, then both the series (19) and (20) should yield the same result. That is,

$$\frac{1}{a_{p-2}} = \frac{1}{p} - \frac{1}{a_p} \qquad \text{or} \qquad \frac{1}{a_{p-2}} + \frac{1}{a_p} = \frac{1}{p}, \tag{21}$$

is the criterion that must be satisfied for the end-correction (*antyasaṃskāra*) to lead to the exact result.

The criterion given by (21) is trivially satisfied when we choose $a_{p-2} = a_p = 2p$. However, this value $2p$ cannot be assigned to both the correction-divisors[4] a_{p-2} and a_p because both the corrections should follow the same rule. That is,

$$\begin{array}{llcl} \text{if} & a_{p-2} = 2p & \Rightarrow & a_p = 2(p+2) \\ \text{or, if} & a_p = 2p & \Rightarrow & a_{p-2} = 2(p-2). \end{array}$$

We can, however, have both a_{p-2} and a_p as close to $2p$ as possible. The choice of $a_{p-2} = 2p - 2$ and $a_p = 2p + 2$ would fulfill the above criteria. It may also be noted that there will persist a difference of 4 between a_{p-2} and a_p since $p - 2$ and p are doubled. Hence, the first (order) estimate of the correction-divisor is given as, "double the even number above the last odd-number divisor p",

$$a_p = 2(p+1). \tag{22}$$

But, it can be seen right away that, with this value of the correction-divisor, the condition for accuracy stated above in (21) is not exactly satisfied. Hence a measure of inaccuracy or error called *sthaulya* $E(p)$ is introduced.

$$E(p) = \left[\frac{1}{a_{p-2}} + \frac{1}{a_p}\right] - \frac{1}{p}. \tag{23}$$

Since this error cannot be eliminated, the objective is to find the correction-divisors a_p such that the inaccuracy $E(p)$ is minimised.

When we set $a_p = 2(p+1)$, the inaccuracy will be

$$\begin{aligned} E(p) &= \left[\frac{1}{(2p-2)} + \frac{1}{(2p+2)}\right] - \frac{1}{p} \\ &= \frac{1}{(p^3 - p)}. \end{aligned} \tag{24}$$

[4] By the term correction-divisor (*saṃskāra-hāraka*), the divisor of the end-correction is meant.

This estimate of the inaccuracy, E_p being positive, shows that the correction has been overdone and hence there has to be a reduction in the magnitude of the end-correction chosen. This means that the correction-divisor has to be increased. If we take $a_p = 2p + 3$, thereby leading to $a_{p-2} = 2p - 1$, we have

$$\begin{aligned} E(p) &= \left[\frac{1}{(2p-1)} + \frac{1}{(2p+3)}\right] - \frac{1}{p} \\ &= \frac{(-2p+3)}{(4p^3+4p^2-3p)} . \end{aligned} \tag{25}$$

Now, the inaccuracy happens to be negative. But, more importantly, it has a term proportional to p in the numerator. Hence, for large p, $E(p)$ given by (25) varies inversely as p^2, while for the divisor given by (22), $E(p)$ as given by (24) varied inversely as p^3. In fact, it can be shown that among all possible correction divisors of the type $a_p = 2p + m$, where m is an integer, the choice of $m = 2$ is optimal, as in all other cases there will arise a term proportional to p in the numerator of the inaccuracy $E(p)$.

From (24) and (25) it is obvious that, if we want to reduce the inaccuracy and thereby obtain a better correction, then a number less than 1 has to be added to the correction-divisor (22) given above. If we try adding *rūpa* (unity) divided by the correction divisor itself, i.e., if we set $a_p = 2p + 2 + \frac{1}{(2p+2)}$, the contributions from the correction-divisors get multiplied essentially by $\left(\frac{1}{2p}\right)$. Hence, to get rid of the higher order contributions, we need an extra factor of 4, which will be achieved if we take the correction divisor to be

$$a_p = (2p+2) + \frac{4}{(2p+2)} = \frac{(2p+2)^2+4}{(2p+2)} . \tag{26}$$

Then, correspondingly, we have

$$a_{p-2} = (2p-2) + \frac{4}{(2p-2)} = \frac{(2p-2)^2+4}{(2p-2)} . \tag{27}$$

We can then calculate the inaccuracy to be

$$E(p) = \left[\frac{1}{(2p-2) + \dfrac{4}{2p-2}} + \frac{1}{(2p+2) + \dfrac{4}{2p+2}}\right] - \left(\frac{1}{p}\right)$$

$$= \left[\frac{(4p^3)}{(4p^4+16)}\right] - \frac{(16p^4+64)}{4p(4p^4+16)}$$
$$= \frac{-4}{(p^5+4p)}. \tag{28}$$

Clearly, the *sthaulya* with this (second order) correction divisor has improved considerably, in that it is now proportional to the inverse fifth power of the odd number.[5]

At this stage, we may display the result obtained for the circumference with the correction term as follows. If only the first order correction (22) is employed, then we will have

$$C \approx 4D\left[1 - \frac{1}{3} + \ldots + (-1)^{\frac{(p-1)}{2}}\frac{1}{p} + (-1)^{\frac{(p+1)}{2}}\frac{1}{(2p+2)}\right]. \tag{29}$$

If the second order correction (23) is taken into account, we have

$$\begin{aligned} C &\approx 4D\left[1 - \frac{1}{3} + \ldots + (-1)^{\frac{(p-1)}{2}}\frac{1}{p} + (-1)^{\frac{(p+1)}{2}}\frac{1}{(2p+2) + \frac{4}{(2p+2)}}\right] \\ &\approx 4D\left[1 - \frac{1}{3} + \ldots + (-1)^{\frac{(p-1)}{2}}\frac{1}{p} + (-1)^{\frac{(p+1)}{2}}\frac{\frac{(p+1)}{2}}{(p+1)^2+1}\right]. \end{aligned} \tag{30}$$

The verse due to Mādhava presenting the infinite series for $\frac{\pi}{4}$ cited earlier is in fact, the first of a group of four verses that present the series along with the above end-correction. The other verse presenting the end-correction represented by (16) is given by [8, p. 390]

> *ante samasaṅkhyādalavargaḥ saiko guṇaḥ sa eva punaḥ |*
> *yugaguṇito rūpayutaḥ samasaṃkhyādalahato bhaved hāraḥ ||*
>
> *triśarādiviṣamasaṅkhyāharaṇāt parametadeva vā kāryam ||*
>
> Square of half of the even term added by unity is [taken to be] the multiplier. Again, the same multiplied by four, added by

[5] It may be noted that if we take any other correction-divisor $a_p = 2p+2+\frac{m}{(2p+2)}$, where m is an integer, we will end up having a contribution proportional to p^2 in the numerator of the inaccuracy $E(p)$, unless $m = 4$. Thus the above form (26) is the optimal second order choice for the correction-divisor.

> one and [the sum further] multiplied by half of the even number becomes the divisor. Instead of keeping on dividing by three, five, etc., it is better to apply this [end-correction].

Incorporating the end-correction given by the above verse, Mādhava series takes the form

$$C \approx 4D\Bigg[1 - \frac{1}{3} + \ldots + (-1)^{\frac{(p-1)}{2}}\frac{1}{p} + \\ + (-1)^{\frac{(p+1)}{2}} \frac{\left(\frac{p+1}{2}\right)^2 + 1}{\left(\left(\left(\frac{p+1}{2}\right)^2 + 1\right)4 + 1\right)\left(\frac{p+1}{2}\right)}\Bigg]. \tag{31}$$

Graph depicting the variation of error in the estimate of π using the three successive end-corrections given by (14)–(16) by truncating the series at different values of p, is shown in Figure 2. For the purpose of convenience, in the figure, we have referred to the plots corresponding to the three successive end-correction terms as first, second and third order respectively. It may be noted that, when we use the III order end-correction, by just considering about 25 terms in the series, we are able to obtain π value correct to 10 decimal places. For more detailed treatment on the topic the readers are referred to *Yuktibhāṣā* [6, pp. 201-207].

5 Circumference by an alternative method

Among the various forms in which the circumference of a circle has been expressed by Mādhava, the following verses present it in a specific form, the rationale behind which—as will be shown below—can be easily understood (by those acquainted with modern calculus) using the Taylor series expansion of the arc-tangent function. The series given by Mādhava is

$$C = \sqrt{12D^2} - \frac{\sqrt{12D^2}}{3.3} + \frac{\sqrt{12D^2}}{3^2.5} - \frac{\sqrt{12D^2}}{3^3.7} + \cdots, \tag{32}$$

which is encoded in the form of verses as follows [8, p. 386]:

> *vyāsavargād ravihatāt padaṃ syāt prathamaṃ phalam |*
> *tadāditastrisaṅkhyaaptaṃ phalaṃ syāduttarottaram ||*

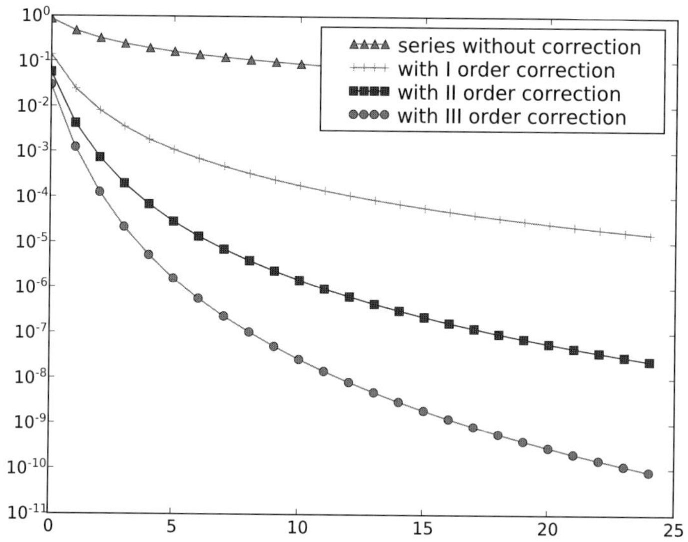

Figure 2: Graph depicting the accuracy that is obtained in making an estimate of the value of π using the Mādhava series by truncating it at different values of p and employing the three successive end-corrections given by (14)–(16). Here x-axis represents the number of terms considered in the series which is $\frac{p+1}{2}$.

rūpādyayugmasaṅkhyābhirhṛteṣveṣu yathākramam |
viṣamāṇāṃ yutestyaktavā samā hi paridhirbhavet ||

The square-root of the product of the square of the diameter and 12 (*ravi*) is the first result (*prathamaphala*). Divide that, and all the successive *phalas* obtained [from the previous one] by three. [Keep generating *phalas*] as you desire (*yāvadicchan*). Then these [*phalas*] are divided, in order, by the odd numbers 1,3, etc. When the sum of the even ones is subtracted from that of the odd ones, the circumference is obtained.

The procedure for obtaining the circumference given in the verse *vyāsavargāt*... may be outlined as follows. Here, the text prescribes the procedure to find several *phalas* P_1, P_2, P_3 and so on upto P_n where n is an integer of our choice (*yāvadicchan*). The successive *phalas* are to be obtained from the preceding *phalas* by dividing by three, viz., $P_{i+1} = \frac{P_i}{3}$, with the first *phala*, P_1, defined as

$$\begin{aligned} P_1 &= \sqrt{ravi \times \textit{vyāsavarga}} \\ &= \sqrt{12D^2}, \end{aligned} \tag{33}$$

where D is the diameter of the circle. It is further said that the *phalas* P_i $(i = 1, 2, \ldots n)$ thus obtained have to be divided by the odd numbers 1,3,5, etc. sequentially. That is,

$$P_{i+1} = \frac{P_1}{3^i.(2i+1)} \qquad (i = 1, 2, 3 \ldots). \tag{34}$$

Now, the circumference is said to be obtained by subtracting the sum of the odd *phalas* from that of the even ones. In other words, if C is the circumference, then

$$\begin{aligned} C &= \sum_{i=1}^{n} (-1)^{i+1} P_i \\ &= \sqrt{12D^2} - \frac{\sqrt{12D^2}}{3.3} + \frac{\sqrt{12D^2}}{3^2.5} - \cdots. \end{aligned} \tag{35}$$

The rationale behind the above expression can be understood as follows. It is well known that the Taylor expansion of $\frac{1}{1+x^2}$ is

$$\frac{1}{1+x^2} = 1 - x^2 + x^4 - x^6 + \ldots \tag{36}$$

Since the derivative of $\tan^{-1} x = \frac{1}{1+x^2}$, by integrating the above equation we get the Taylor series expansion of $\tan^{-1} x$. That is,

$$\tan^{-1} x = x - \frac{x^3}{3} + \frac{x^5}{5} - \frac{x^7}{7} + \ldots \tag{37}$$

By making the substitution $\theta = tan^{-1}x$, we have $x = \tan\theta = \frac{\sin\theta}{\cos\theta}$. Hence, the above equation reduces to the form

$$\begin{aligned} \theta &= \left(\frac{\sin\theta}{\cos\theta}\right) - \frac{1}{3}\left(\frac{\sin\theta}{\cos\theta}\right)^3 + \frac{1}{5}\left(\frac{\sin\theta}{\cos\theta}\right)^5 - \ldots \\ \text{or } R\theta &= \frac{R\sin\theta}{1\cos\theta} - \frac{R\sin\theta}{3\cos\theta}\left(\frac{\sin\theta}{\cos\theta}\right)^2 + \frac{R\sin\theta}{5\cos\theta}\left(\frac{\sin\theta}{\cos\theta}\right)^4 - \ldots \end{aligned} \tag{38}$$

If we take the arc-length equal to one-twelfth of the circumference, i.e., ($R\theta = \frac{C}{12}$), which is equivalent to taking $\theta = \frac{\pi}{6}$, then $\frac{\sin\theta}{\cos\theta} = \frac{1}{\sqrt{3}}$. Substituting this in the above series we get

$$\begin{aligned}
\frac{C}{12} &= R\left(\frac{1}{\sqrt{3}}\right) - \frac{R}{3}\left(\frac{1}{3\sqrt{3}}\right) + \frac{R}{5}\left(\frac{1}{3^2\sqrt{3}}\right) \\
\frac{C}{12} &= \left(\frac{R}{\sqrt{3}}\right)\left[1 - \frac{1}{3.3} + \frac{1}{5.3^2} - \frac{1}{7.3^3}\cdots\right] \\
C &= 12\left(\frac{D}{\sqrt{12}}\right)\left[1 - \frac{1}{3.3} + \frac{1}{3^2.5} - \frac{1}{3^3.7}\cdots\right] \\
&= \sqrt{12D^2}\left(1 - \frac{1}{3.3} + \frac{1}{3^2.5} - \frac{1}{3^3.7}\cdots\right), \qquad (39)
\end{aligned}$$

which is the same as the series given by (35).

6 Fast convergent approximations to Mādhava series

Consider the expressions for the circumference of a circle given by (6*b*) and (39) in terms of its diameter. It may be noted that in the former series only the successive odd numbers appear in the denominator with first power, whereas, in the latter, we find successive powers of 3 (3^i, $i = 0, 1, 2, \ldots$) multiplying the odd numbers appearing in the denominator. Obviously the latter series will be converging faster and hence would be fetching far more accurate results by considering fewer terms in the series.

Besides this series, which of course is unquestionably a calculus masterpiece, Mādhava seems to have given many interesting fast convergent approximations of his series by incorporating the end-corrections given by (14)–(16) into the series itself. In what follows, we present some of the transformed versions of the Mādhava series presented by Śaṅkara Vāriyar in his *Kriyākramakarī* [8, p. 390].

6.1 Transformed Mādhava series: I

samapañcāhatayo yā rūpādyayujāṃ caturghnamūlayutāḥ |
tābhiḥ ṣoḍaśaguṇitād vyāsāt pṛthagāhṛte tu viṣamayuteḥ |
samaphalayoge tyakte syādiṣṭavyāsasambhavaḥ paridhiḥ ||

The fifth-powers of odd numbers commencing from one are added to four times themselves. By dividing the product of six-

> teen and the diameter by these separately, and subtracting the sum of the even terms from the odd ones we get the circumference corresponding to a given diameter.

The expression for the circumference of a circle presented in the above verses may be written as

$$\begin{aligned} C &= \frac{16D}{1^5+4.1} - \frac{16D}{3^5+4.3} + \frac{16D}{5^5+4.5} - \cdots \\ &= 16D\left(\frac{1}{1^5+4.1} - \frac{1}{3^5+4.3} + \frac{1}{5^5+4.5} - \cdots\right) \end{aligned} \tag{40}$$

The rationale behind the above expression can be understood with the help of equations (18) and (23) giving the expression for circumference and the expression for inaccuracy of *sthaulya* respectively. Using them, the circumference may be expressed in terms of *sthaulyas* as follows:

$$\begin{aligned} C &= 4D\left[\left(1-\frac{1}{a_1}\right) + \left(\frac{1}{a_1}+\frac{1}{a_3}-\frac{1}{3}\right) - \left(\frac{1}{a_3}+\frac{1}{a_5}-\frac{1}{5}\right) - \cdots\right] \\ &= 4D\left[\left(1-\frac{1}{a_1}\right) + E(3) - E(5) + E(7) - \cdots\right], \end{aligned} \tag{41}$$

Again using (28) and (15) in (41), we get

$$\begin{aligned} C &= 4D\left(1-\frac{1}{5}\right) - 16D\left[\frac{1}{(3^5+4.3)} - \frac{1}{(5^5+4.5)} + \frac{1}{(7^5+4.7)} - \cdots\right] \\ &= 16D\left[\frac{1}{(1^5+4.1)} - \frac{1}{(3^5+4.3)} + \frac{1}{(5^5+4.5)} - \frac{1}{(7^5+4.7)} + \cdots\right], \end{aligned} \tag{42}$$

which is the same as the expression (40) given in the above verse. Further, it may be noted that each term in the above transformed Mādhava series involves fifth power of the odd numbers in the denominator, whereas the original Mādhava series had only the first power of odd numbers appearing in the denominator. Thus it goes without saying that the transformed Mādhava series given by (42) would be converging much faster than the original series (6*b*).

6.2 Transformed Mādhava series: II

> *vyāsādvāridhinihatāt pṛthagāptaṃ tryādyayugvimūlaghanaiḥ |*
> *trighnavyāse svamṛṇaṃ kramaśaḥ kṛtvāpi paridhirāneyaḥ ||*
>
> The diameter multiplied by four is divided separately by the cubes of the odd numbers starting with three diminished by their bases. By applying this positively and negatively, in order, to the diameter multiplied by three, the circumference should be obtained.

By choosing $a_p = 2p + 2$ and $a_{p-2} = 2p - 2$ and substituting them in (41), we get a transformed series

$$\begin{aligned} C &= 3D + \frac{4D}{(3^3-3)} - \frac{4D}{(5^3-5)} + \frac{4D}{(7^3-7)} - \cdots \\ &= 4D\left[\frac{3}{4} + \frac{1}{(3^3-3)} - \frac{1}{(5^3-5)} + \frac{1}{(7^3-7)} - \cdots\right], \end{aligned} \tag{43}$$

which again would have faster convergence since the cubes of odd numbers appear in the denominator unlike the first power in the original series. The rate of convergence of the above series in comparison with the series given by (42) is shown in Figure 3. In this figure, as well as in Figures 4 and 5, the series depicted by the graphs are indicated by the beginning of the verse like '*samapañcāhati*' etc., that presents the series.

6.3 Transformed Mādhava series: III

> *dvyādiyujāṃ vā kṛtayaḥ vyekā hārā dvinighnaviṣkambhe |*
> *dhanamṛṇamante'ntyordhvagataujakṛtirdvisahitā*[6] *haro dvighnī ||*
>
> Or, the squares of the even numbers starting with two, diminished by one are considered as divisors (*hāras*) [of the product of four and the diameter]. These have to be added to and subtracted from twice the diameter [alternatively]. The square of the odd number that succeeds the last even number, added by two and doubled would be the [correction] divisor.

[6]The reading found in *Kriyākramakarī* is '*kṛtidvisahitā*'. However, it is grammatically incorrect if the *visarga* is missing in *kṛtiḥ*—which actually takes the form of '*r*' due to *sandhi* rules. Hence, we have included the correct form in the main text by introducing *visarga*.

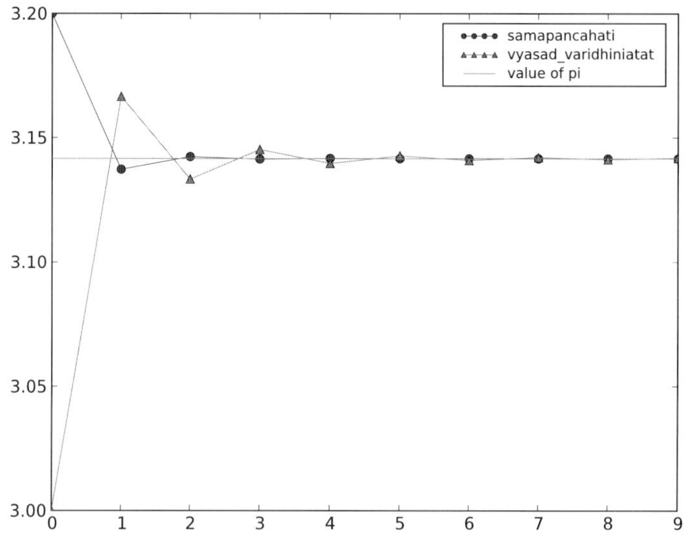

Figure 3: Graph depicting the convergence of the series given by (42) and (43). Here x-axis represents the number of terms considered in the series.

Since the above verse is composed in terse style, a brief explanation would be in place to facilitate the understanding of its content. Here, the author simply specifies the denominators of the various terms that would constitute the series, without specifying the numerators. This, however, cannot be considered as a lapse or omission on the part of the author, since he simply follows an accepted style of composition in Sanskrit. The numerators can be easily understood to be $4D$, by a process known as *anuṣaṅga* (implicit connection), from the previous verse, that commences with '*vyāsādvāridhinihatāt pṛthagāptam*' which means 'the product of four and diameter separately divided by'.

The denominators of the terms that constitute this series are stated to be squares of the even numbers (starting with two) and diminished by one. Thus, the terms—leaving the first one—constituting the present series may be written as

$$\frac{4D}{(2i)^2 - 1} \qquad (i = 1, 2, \ldots). \tag{44}$$

It is further said that these terms have to be applied to $2D$ positively and negatively, alternatively. Hence, the resulting series would be

$$\begin{aligned} C &= 2D + \frac{4D}{(2^2-1)} - \frac{4D}{(4^2-1)} + \frac{4D}{(6^2-1)} - \cdots \\ &= 4D\left[\frac{1}{2} + \frac{1}{(2^2-1)} - \frac{1}{(4^2-1)} + \frac{1}{(6^2-1)} - \cdots\right] \end{aligned} \tag{45}$$

Here again, the rationale behind the above series can be understood by choosing a_p and a_{p-2} as $2p$ and $2p-4$ respectively and substituting in (41). Doing so, we get

$$\begin{aligned} C &= 4D\left[\frac{1}{2} + \frac{1}{3} - \frac{1}{15} + \frac{1}{35} - \cdots\right] \\ &= 4D\left[\frac{1}{2} + \frac{1}{(2^2-1)} - \frac{1}{(4^2-1)} + \frac{1}{(6^2-1)} - \cdots\right], \end{aligned} \tag{46}$$

which is same as the series given in the verse. Having given the series, in the latter half of the verse it is mentioned that the above series could be terminated by applying a correction-term after dividing by n^2-1, where n is an even number of our choice. The resulting series along with the end-correction would be of the form

$$\begin{aligned} C \approx 4D\Bigg[&\frac{1}{2} + \frac{1}{(2^2-1)} - \ldots + (-1)^{\frac{n-2}{2}}\frac{1}{(n^2-1)} + \\ &+ (-1)^{\frac{n}{2}}\frac{1}{2[(n+1)^2+2]}\Bigg]. \end{aligned} \tag{47}$$

It can be easily seen from Figure 4 that the inclusion of the correction-term significantly increases the rate of convergence of the series given by (46).

6.4 Transformed Mādhava series: IV

dvyādeścaturāderva caturadhikānāṃ nirekavargāssyuḥ |
hārāḥ kuñjaraguṇito viṣkambhaḥ svamatikalpito bhājyaḥ |
phalayutirekatra vṛtirbhājyadalaṃ phalavihīnamanyatra ||

The squares of the numbers starting from two or four and increased by four, are diminished by unity. If these [numbers obtained sequentially] are taken to be the divisors (*hāras*) then the desired diameter multiplied by eight will be dividend (*bhājya*).

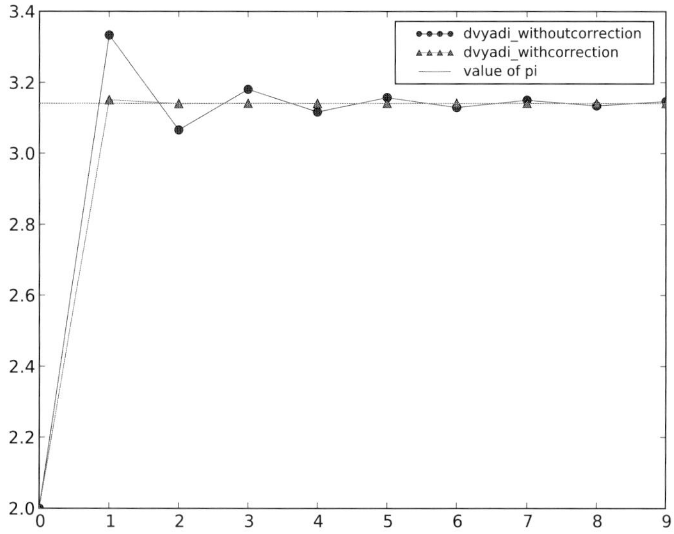

Figure 4: Graph depicting the convergence of the series given by (46) and (47). Here x-axis indicates the number of terms considered in the series.

> [Then] the circumference would be the sum of the results at one place [where it is started with the square of two] and half of the *bhājya* diminished by [sum of] all the terms at the other place.

The series given by (46) is an alternating series. Mādhava by grouping the terms in the series, seems to have split this into two series, and has represented them as

$$C = 8D\left[\frac{1}{(2^2-1)} + \frac{1}{(6^2-1)} + \frac{1}{(10^2-1)} + \cdots\right]. \tag{48}$$

Also, $$C = 8D\left[\frac{1}{2} - \frac{1}{(4^2-1)} - \frac{1}{(8^2-1)} - \frac{1}{(12^2-1)} - \cdots\right]. \tag{49}$$

It can be easily seen that the sum of (48) and (49) gives the series (46). Another feature that is noteworthy of (48) and (49) is that, they are monotonically increasing and decreasing respectively (see Figure 5). In fact, both of them asymptotically approach the value of π as shown in the figure.

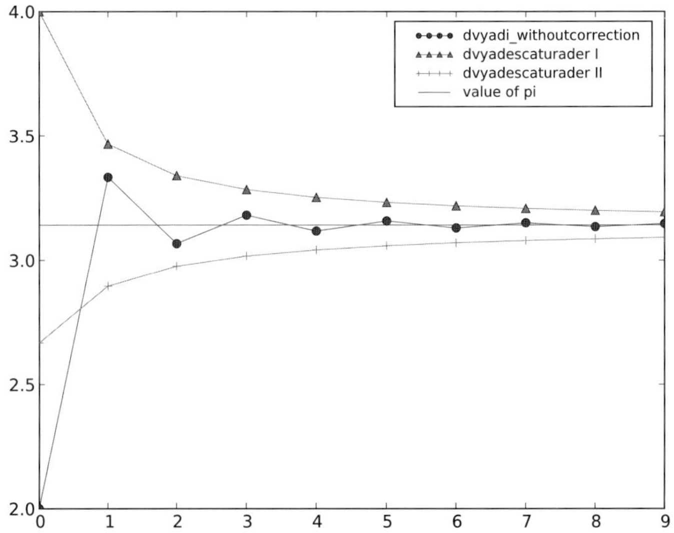

Figure 5: Graph depicting the convergence of the series given by (46) and its variant forms given by (48) and (49). Here x-axis represents number of terms considered in the series.

Finally, before concluding the paper, we would like to present a glimpse of the interesting discussion on the irrationality of π, found in monumental work *Āryabhaṭīyabhāṣya*—the magnum opus of Nīlakaṇṭha (c. 1500 CE).

7. Irrationality of π

Having specified the ratio of the circumference to the diameter of a circle, Āryabhaṭa in his *Āryabhaṭīya* (c. 499 CE) refers to the value[7] as 'approximate' (*āsanna*). Nīlakaṇṭha while commenting upon the verse raises the question: "Why then has an approximate value been mentioned here instead of the actual value?", and then explains [1, p. 41]:

> Given a certain unit of measurement in terms of which the diameter (*vyāsa*) specified has no [fractional] part (*niravayava*), the same measure when employed to specify the circumference

[7]The value given is $\frac{62832}{20000} = 3.1416$, correct to four decimal places.

> (*paridhi*) will certainly have a [fractional] part (*sāvayava*). ... Even if you go a long way (i.e., keep on reducing the measure of the unit employed), the fractional part [in specifying one of them] will only become very small. A situation in which there will be no [fractional] part is impossible, and this is what is the import [of the expression *āsanna*].

Evidently, what Nīlakaṇṭha is trying to explain here is the incommensurability of the circumference and the diameter of a circle. Particularly, the last line of the above quote—where Nīlakaṇṭha in no uncertain terms mentions that, however small you may choose your unit of measurement to be, the two quantities will never become commensurate—is noteworthy.

8. Conclusion

It should be evident from the above discussion, that the notion that mathematics in India has been simply a handmaiden to astronomy seems to be misconcieved. For, it were to be so, then it may be difficult to explain as to why Nīlakaṇṭha cogitated on the irrationality of π, or Mādhava evolved general methods that in principle enables one to obtain values of π that are correct to as many number of decimal places as one desires. More detailed studies are required to understand the motivation behind the Kerala astronomers developing techniques for arriving at very accurate π values (upto 17 decimal places).

Also it has been well-argued by Raju, in one of his recent publications [10], that the historians of the past have failed to understand and appreciate the distinct approach taken by Indians to mathematics. In fact, on occasions—either due to ignorance or misunderstanding or for reasons not evident to us—the historians have been quite dismissive regarding the Indian contributions as well [3,9]. But as demonstrated earlier, the contributions of the Kerala school of astronomers and mathematicians to the development of 'what may be called classical analysis'—in the context of finding the relationship between the circumference and the diameter of a circle—have been quite significant, and their methodology needs to be studied in depth for its sheer pedagogical implications, if not for anything else.

Acknowledgements

The authors would like to express their gratitude to Sir Dorabji Tata Trust, Mumbai and the National Academy of Sciences, India, for the support received from them by way of sanctioning projects to carry out research on scientific manuscripts in Sanskrit.

References

[1] *Āryabhaṭīya* of Āryabhaṭa with *Āryabhaṭīyabhāṣya* by Nīlakaṇṭha Somayājī, Ed. by K. Sāmbaśiva Śāstrī, Trivandrum 1930.

[2] Boyer C B, *History of Mathematics*, John Wiley & Sons, New York 1968.

[3] Boyer C B, *The History of the Calculus and its Conceptual Development*, Dover, New York 1949, pp. 61–62: The Hindus apparently were attracted by the arithmetical and computational aspects of mathematics, rather than by the geometrical and rational features of the subject which had appealed so strongly to the Hellenic mind. Their name for mathematics, *gaṇita*, meaning literally the 'science of calculation' well characterizes this preference. They delighted more in the tricks that could be played with numbers than in the thoughts the mind could produce, so that neither Euclidean geometry nor Aristotelian logic made a strong impression upon them. The Pythagorean problem of the incommensurable, which was of intense interest to Greek geometers, was of little import to Hindu mathematicians, who treated rational and irrational quantities, curvilinear and rectilinear magnitudes indiscriminately.

[4] Roger Cooke, *The History of Mathematics*, John Wiley & Sons, New York 1997.

[5] Edwards C H, *The Historical Development of the Calculus*, Springer, New York 1979.

[6] *Gaṇita-yukti-bhāṣā* of Jyeṣṭhadeva, Ed. and Tr. by K. V. Sarma, with Explanatory Notes by K. Ramasubramanian, M. D. Srinivas and M. S. Sriram, 2 Vols, Hindustan Book Agency, New Delhi 2008. Reprint Springer Verlag, 2009.

[7] Joseph Needham, *The Shorter Science and Civilization in China*, Colin A. Ronan, CUP, 1981, vol. 2, p. 23.

[8] *Līlāvatī* of Bhāskarācārya with *Kriyākramakarī* of Śaṅkara Vāriyar, Ed. by K. V. Sarma, VVRI, Hoshiarpur 1975.

[9] Morris Kline, *Mathematical Thought from Ancient to Modern Times*, Oxford 1972, p. 190: (With the Hindus) there is much good procedure and technical facility, but no evidence that they considered proof at all. They had rules, but apparenly no logical scruples ...

[10] Raju C K, *Cultural Foundations of Mathematics: The Nature of Mathematical Proof and the Transmission of Calculus from India to Europe in the 16th c.* CE, PHISPC, Vol. X, Part 4, Pearson Longmann, New Delhi 2007.

[11] Ramasubramanian K, Srinivas M D, 'Development of Calculus in India', *Studies in the History of Mathematics*, ed. by C. S. Seshadri, Hindustan Book Agency, New Delhi 2010, pp. 201–86.

[12] Ranjan Roy, 'The Discovery of the Series formula for π by *Leibniz, Gregory* and Nīlakaṇṭ

PRINTED ON PERMANENT PAPER • IMPRIME SUR PAPIER PERMANENT • GEDRUKT OP DUURZAAM PAPIER - ISO 9706

N.V. PEETERS S.A., WAROTSTRAAT 50, B-3020 HERENT